Uncertainty and Ground Conditions

To Annelies, only she knows why.

Uncertainty and Ground Conditions: A Risk Management Approach

Martin van Staveren

AMSTERDAM • BOSTON • HEIDELBERG • LONDON • NEW YORK • OXFORD
PARIS • SAN DIEGO • SAN FRANCISCO • SINGAPORE • SYDNEY • TOKYO

ELSEVIER

Butterworth-Heinemann is an imprint of Elsevier

Butterworth-Heinemann is an imprint of Elsevier
Linacre House, Jordan Hill, Oxford OX2 8DP, UK
30 Corporate Drive, Suite 400, Burlington, MA 01803, USA

First edition 2006

British Library Cataloging in Publication Data
A catalog record for this book is available from the British Library

Library of Congress Cataloguing in Publication Data
A catalogue record for this book is available from the Library of Congress

ISBN–13: 978-0-75-066958-0
ISBN–10: 0-75-066958-6

For information on all Butterworth-Heinemann publications please visit our
website at www.books.elsevier.com

Printed and bound in the Great Britain

Contents

vii

Acknowledgements

The open GeoQ framework for managing ground-related risk, as presented in this book, emerged after the start of the third millennium. Writing this book has only been possible with the indispensable support of many pioneering individuals and teams over the last years, in The Netherlands and abroad. These professionals have both technical and non-technical backgrounds and work at many different firms, government agencies, universities and institutions. They dared to try-out parts of the GeoQ concept in their projects and together we learned a lot from it. These experiments resulted in the case studies in this book. I want to thank all of these professionals, for their opinions, suggestions and feedback on the GeoQ risk management approach.

Special thanks go to the many organizations who gave me opportunities to present the GeoQ process and to receive their feedback, including the Delft University of Technology, Rijkswaterstaat of the Dutch Ministry of Public Works and Water Management, Fugro, Heijmans, Volker Wessels, CROW, CUR, Norwegian Geotechnical Institute, Geo Research Institute in Japan, GeoHohai of Hohai University in China and Argonne National Laboratory in the USA.

There are a number of people I want to mention in particular. The late David Price, the first professor of Engineering Geology at the Delft University of Technology, is my great inspirator in engineering geology, particularly concerning the relationship between risk, ground investigations and contractual aspects.

Robert J. Smith of Wickwire Gavin in Madison and Red Robinson of Shannon and Wilson in Seattle shared their viable GBR experiences and information with me during my visit in December 2000. With Alan Pace I had interesting exchanges of opinions and information about ground risk allocation. Thanks go to all three of them.

Joop Halman, professor at the University of Twente, thank you for sharing some knowledge and experiences with risk management in general and the Risk Diagnosing Method for industrial innovation management in particular.

Joost Wentink, managing director of GeoDelft and Erik Janse, GeoDelft's operational director, thanks for providing me some appreciated time to work on this book. Erik issued me with interesting information as well. Colleagues of GeoDelft's Management Team, I appreciate your understanding for my ground risk management bias and drive to complete this book in due time. In particular, Marco Hutteman, I remember the inspiring management development week we shared at Pierrefitte in France, autumn 2005, which proved to be a catalyst for this book. Gerard Vennegoor of Pierrefitte, a thanks to you for your wise words.

Furthermore, thanks to all my colleagues at GeoDelft, the pioneers from the GeoQ-team and especially the teams of the GeoLab Department. These teams proved to have an exceptional degree of self-management during the periods I worked on the book. Jan Mul of the GeoDelft library, many thanks for the many papers and references you provided over the years. Alex Hollingsworth, Lanh Te and Jackie Holding of Elsevier in Oxford, I really appreciate your professional advice and support.

Finally, above all, Annelies, Charlotte, Josephine and Frédérique, many thanks for all the support and understanding, which made it possible to transform my challenge into this book.

PART ONE

The context of ground risk management in the construction industry

1 Introduction

A new type of ground risk management book

How can we live without construction? It fulfils many of our fundamental needs and has existed since the earliest development of mankind. Incorporating engineering and maintenance activities, *construction*, provides us with houses, schools, hospitals, industrial plants and infrastructure. We are all affected by these structures, hour after hour, day after day, year after year.

There is no construction without *ground*. Any kind of construction needs a foundation. Any construction, whether very small or extremely large, has some form of connection with the inherently uncertain ground. Our ability to cope with this uncertainty will make a difference between our foundation settlements or not, between excess groundwater in our basements or not, or even whether our structures collapse during an earthquake or not.

Until now, the ground has always been a major driver of risk in many construction projects all over the world. This is reflected in the relatively high failure costs and often small profit margins in the construction industry. Many projects are completed at a higher cost than estimated, as well as much later than scheduled. This causes serious additional expenditure for clients, reduced profitability or even losses for contractors and a lot of irritation for the public.

For many years, *risk management* has added value in many sectors and industries, such as the financial sector, the chemical industry and the offshore industry. In construction, however, risk management has not been entirely incorporated and exploited, in spite of the industry's inherent uncertainties and high risks. The application of well-structured risk management during all project stages, from feasibility through to construction and maintenance, needs to be started or extended to many more projects. This situation is particularly apparent in ground-related engineering and construction activities.

A serious obstruction to the introduction and application of risk management is the *people factor*. Together, we *are* that people factor. Typical human attitudes and behaviour, driven by unawareness and fear, often prevent us from considering risk in a timely and effective way. As a result, we will miss opportunities to optimize projects and benefits for our organizations, our clients and our societies as a whole remain hidden and untouched.

The combination of these four interrelated aspects, *construction*, *ground*, *risk management* and the *people factor*, provides an opportunity for a new type of risk management book. Is there a need for it? Yes, I think there certainly is, in spite of a number of related books published over recent years. Examples are those written by Edwards and Bowen (2005), Weatherhead et al. (2005), Smith (2003, 1998), Boothroyd and Emmet (1996), Godfrey (1996), Edwards (1995), Flanagan and Norman (1993) and Thompson and Perry (1992). All these books cover risk management in the construction industry, but do not focus on ground risk management. The number of available books that cover ground-related risk management in particular is limited. Although works by Clayton (2001), Hatem (1998) and Skipp (1993) do focus on ground risk management, they pay little attention to the people factor. None of these books combines the four interrelated factors dealt with in this book.

Objectives and target readerships

The main objective of this book is to contribute to the application of cost-effective ground risk management. It considers ground conditions in their widest definition and includes all types of ground, groundwater, ground-related pollution, and all forms of man-made structure. The latter refers to buried structures, such as pipelines, piles or archaeological remains.

In today's increasingly global market we must differentiate or die, according to Trout and Rivkin (2000) in their guideline on how to survive killer-competition. Ground-related innovations in engineering and construction are urgently required to gain competitive advantage. This book's secondary objective is therefore that ground risk management should act as a sort of airbag against the inherent business risks of innovations. A similar risk management approach has been used in other industries. For instance, the Risk Diagnosing Method (RDM) proposed by Keizer, Halman and Song (2002) has been successfully applied in the consumer electronics and food industries.

GeoQ, where Q stands for quality, will become the vehicle to meet our objectives. It is an easy-to-use and flexible framework for ground risk management during the entire life cycle of all types of construction projects. It is independent of the type of ground conditions expected and can reveal many hidden and ground-related opportunities, such as cost savings, tighter schedules, improved project quality and increased profitability for a lot of stakeholders. Anyone can make GeoQ fit-for-purpose, to meet the specific requirements of any small or large construction project, anywhere in the world.

Given these objectives, the main target readership will include civil engineering and construction professionals involved in ground-related issues in some way. They may be working with contractors, engineering firms and clients, studying for BSc, MSc and MBA degrees or teaching and performing research at universities and institutes. Here we recognize construction managers, project planners, project designers, geotechnical engineers, soil engineers, rock engineers, engineering geologists, ground-related scientists, graduate and postgraduate students.

I hope to inspire and motivate this anticipated variety of readers, who will all encounter their ground risks in some form throughout their careers. If many of you start to participate in the worldwide adoption of structured ground risk management, we will be able to make a difference in the rapidly changing construction industry.

State-of-the-art of ground risk management

The state-of-the-art of ground risk management, as presented in this book, is a mixture of theory and practice. It is derived from a variety of engineering, business administration and human sciences and includes many aspects of ground engineering and construction, some physics, statistics and geology, as well as several fundamentals of psychology, sociology, and even some philosophy. According to a modern risk management approach, as proposed by Edwards and Bowen (2005) for instance, risk is considered to form both an obstruction and an opportunity for project success.

Empirical developments are major drivers for innovation, particularly for ground-related engineering and construction. I, therefore, do not intend to present a new scientific risk management theory, but will present a structured and risk-prone way of thinking and doing.

GeoQ ground risk management is a form of process innovation that typically emerged by trial and error. It has been applied in a wide range of projects, including tunnels, (rail)roads and even a waste disposal site, resulting in its present state-of-the-art. Common scientific approaches, such as objectivity and the proof of principle by experiment, are used within the limitations of the available experience. I have included abundant references from a variety of disciplines to support and criticize my opinions about and experience of ground risk management. Colleagues from all over the world suggested many of these, others I approached by purpose or just came across.

The GeoQ framework should not be seen as having arrived its final state of development. It has been introduced only recently and there will be ample opportunity for further improvement. Many of the GeoQ supporting practices that are presented, such as scenario analysis, risk identification and classification methods, ground investigations, and the observational method with monitoring, are not new but are readily available to deliver GeoQ support. Some may demand further development to increase their cost-effectiveness and ease for daily use in ground risk management.

The first part of this book pays particular attention to thinking and reflection, while the second part is mainly concerned with learning by doing. This combination will not be able to prevent each and every project crisis from time to time. Risk management is by no means a panacea capable of preventing all risks in ground engineering and construction activities. If we can merely reduce the probability of such risks occurring, as well as their effects, then the objectives of this book will have been achieved.

The book's structure

The structure of this book is designed to help first-time users, who are not yet familiar with risk management, as well as experienced professionals using the book as a reference guide for applied ground risk management.

According to John Naisbitt (1984): 'What happens is that whenever new technology is introduced in society, there must be a counterbalancing human response – that is high touch – or the technology is rejected.' GeoQ ground risk management, with its technological tools, should be appraised as a form of new technology. Experience teaches the importance of giving ample attention to professional attitudes and behaviour. If not, risk management becomes little more than a tick-box exercise and a waste of our precious time and money. This explains the *high tech* and *high touch* approach in this book: to provide fertile

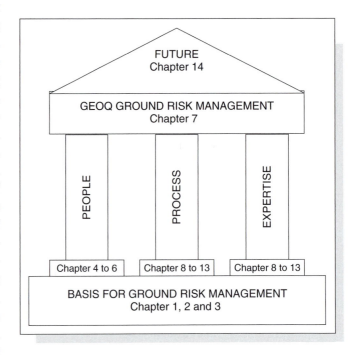

Figure 1.1: The book structure – GeoQ ground risk management with its three pillars.

ground for a wide acceptance and application of ground risk management. Figure 1.1 shows the book's structure, together with the corresponding chapters.

Chapters 1, 2 and 3 serve as the foundation slab and bear the three pillars of GeoQ ground risk management and its future: people, processes and expertise. Following the introduction in Chapter 1, Chapter 2 presents a number of challenges and opportunities for global construction. It serves as an appetiser for Chapter 3, which brings us from uncertainty, via risk, risk management and the ground, to the *concept* of GeoQ.

Chapters 4 to 6 focus on the high touch or human factor, the combination of *people* and *risk*. These chapters highlight the need for risk awareness and the inherent differences in people's perception of risk. Chapter 4 identifies certain characteristics of individual risk perceptions, as well as how individuals can contribute to effective risk management. Chapter 5 explores the interaction of individuals in teams, including aspects such as team culture and risk communication. The way in which teams may contribute to ground risk management is also discussed. Chapter 6 describes how clients and society perceive the risk caused by construction. We can use their insights to guarantee more effective communication about (ground) risk with these stakeholders.

Chapters 7 to 13 explore the high-tech side of ground risk management, in particular the technical-organizational or *process* aspect. These chapters present the *application* of the tried-and-tested GeoQ method in six generic project phases, to provide us and our teams, clients and society with *high quality* construction products. To guarantee maximum benefits, six subsequent risk management steps must be taken in each phase. Chapter 7 introduces this GeoQ process. Chapters 8 through to Chapter 13 present its application during the feasibility, pre-design, design, contracting, construction, and operation and maintenance phases. Each chapter begins with several ground risk mitigation measures and tools, followed by a variety of *case studies*, where GeoQ steps and tools will add value to the project. These are intended to help understand the many types of projects where GeoQ can be applied. Which GeoQ tool should we apply to which situation? There is no generic answer, all that can be said is: *it depends*. Concise summaries are presented at the end of each chapter. Finally, Chapter 14 highlights briefly some of the main opinions and conclusions of this book, followed by some type of outlook to a prosperous construction industry, as perceived from a ground risk management perspective.

The third pillar below the GeoQ concept shown in Figure 1.1 is *expertise* of ground engineering and construction. Many textbooks and papers are available about ground engineering, soil mechanics, rock mechanics, groundwater engineering, environmental engineering and engineering geology and provide plentiful information. This book provides numerous examples of the *benefits* of sound, up-to-date expertise, as well as experience, as these remain ultimately necessary for responsible and cost-effective ground risk management.

In addition to figures and tables, numerous text boxes are included in the chapters. These should be seen as a side step for reflection on the issues presented, whose purpose is to raise awareness and provide fresh insights rather than give definite answers. Readers are invited to jump from chapter to chapter, based on their own interests, experiences and needs. Introductions and summaries in each chapter provide a quick overview of their content. Before starting to apply the GeoQ process using the guidelines and experiences described in Chapters 7 to 13, I recommend that you first read Chapters 4 to 6. After all, it is people like you and me who are responsible for construction and its associated ground risks.

2 *The construction industry – challenges and opportunities*

Introduction

The magnitude of construction is enormous and a number of major challenges affect its current state. This chapter explores the ever increasing *complexity*, the relatively underdeveloped *integrity* and the substantial *failure costs* associated with construction activities. *Ground conditions* play a major role in these aspects.

Is there an ongoing crisis in our industry? Perhaps there is, but the good news is that new solutions and opportunities continue to emerge. In recent years, a number of countries have initiated ambitious change programmes for the construction industry. In this chapter we will meet some of these initiatives, as they may help us cope effectively with the challenges we face.

This chapter introduces the concept of *systems thinking*, a potential key to unlock possibilities for the required industry transformation. It stipulates fertile ground for the concept of *risk management*. The last part of the chapter highlights the need for a critical mass of change-driven professionals. These individuals will be essential for implementing ground-related risk management, as presented in this book, in day-to-day engineering and construction practices. The summary presents the key issues of this chapter.

The magnitude of the world's construction industry

The global construction industry is huge, and will continue to grow substantially in size. Based on a report by Global Insight, Sleight (2005a) predicts an increase in construction spending from 3500 billion US dollars in 2003, to 4800 billion US dollars in 2008 and 6200 billion US dollars in 2013. We should note that a billion is here defined as equal to 1000 million. These figures indicate an enormous worldwide construction market that is expected almost to double in 10 years time.

The USA has the largest construction market, accounting for some 30 per cent of the world's construction activities. Japan is the second largest market with approximately 13 per cent of the market share. In 2003, China was the third largest, representing 7 per cent of the world's construction expenditure. In this top ten of the world's largest construction markets, Germany is in fourth place, followed by France, Italy, the UK, Spain, Canada and The Netherlands, ranked from five to ten. Emerging markets such as India, Brazil and Russia are ranked numbers 11, 13 and 15 respectively. These emerging markets are expected to grow between 3 and 5 per cent per year. Compared with the USA, Japan and Europe, construction in these countries is considered risky. Growth expectations in the USA, Japan and Europe are substantially lower, at approximately 2 per cent per year.

Three main market segments can be distinguished: residential construction, infrastructure and other construction. In the USA, market shares are 44 per cent for residential construction, 32 per cent for infrastructure and 24 per cent for other construction.

It is not only the size of the construction industry that is impressive, its influence on societies is enormous as well, for example as provider of shelter, water supplies and transportation. In many countries, approximately 10 per cent of the work-force is directly involved in construction and its related sectors. The European Construction Technology Platform (ECTP) is an industry-driven initiative whose aim is to act as an umbrella for the construction industry's research initiatives in Europe. Their *Vision-2030 Report* states:

> Europe is facing serious challenges. If we sit back and rest, by 2030, global warming will cause increasing number of disastrous damages from floods and storms. The systems for water supply and wastewater will be dilapidating all over Europe. Workers will continue to die on accidents at work. Traffic congestion problems will reach breaking point and seriously hamper the economic and social development of Europe. Another 10% (sic) the tangible cultural heritage will have been lost. These challenges are for real. It is safe to say that the time for action is now (European Construction Technology Platform, 2005).

Urban issues are of particular importance. Barends and Mischgofsky (2005) predict that without serious concern, research and subsequent actions regarding these issues, the future viability, economic health and quality of life of Europe's citizens will be in jeopardy. Such a situation is not exclusive to Europe, but is also applicable to many other regions. Construction, in its widest sense, must help prevent these negative visions becoming reality.

Water in particular plays a dominant role in construction. The two cannot be separated. Water does seriously hamper our construction activities from time to time and many readers are no doubt familiar with groundwater problems during construction. In addition, a lot of areas in many parts of the world suffer excesses or shortages of water. The construction industry is one of the main sectors able to facilitate long-term solutions and relief. Box 2.1 presents a few thoughts about the relationship between water and construction.

Box 2.1　Water and construction

I was born and raised in The Netherlands, a small country where the majority of the population live well below sea level. Thousands of kilometres of dikes protect the Dutch against attacks by high water. A major flood disaster in 1954 gave rise to the enormous Delta works in the southwestern part of The Netherlands. So far, these flood defences have been able to protect the Dutch. But for how long? The water system in The Netherlands is coming under increasing pressure due to urbanization, as well as a lack of water-catchment areas, land-subsidence and rises in sea-level.

Recent history reminds us how fragile our modern societies are when water conditions become extreme. It is not easy to forget the devastating effects of the tsunami in Southeast Asia in December 2004. Entire populated coastal areas disappeared in Indonesia, Thailand, India and Sri Lanka. More than 250 000 people were killed within just a few hours. In September 2005, the world witnessed the damaging effects of collapsing dikes around New Orleans in the Mississippi delta of the USA, in the wake of the hurricane Katrina. Similar disasters occur all too often in other parts of the world. We are all aware of regular floods in countries such as Bangladesh.

Other parts of the world are confronted with structural droughts, while areas in the Middle East, Africa and China face an increasing shortage of water in the near future. Mega construction projects are needed to bring relief, such as the three huge canal projects in China, for transporting water from major rivers to dry areas.

Smaller projects are also required alongside the mega projects, for example, initiatives by the ICE Commission Engineering Without Frontiers (EWF) to install water pumps and small dams for the provision of clean water in remote villages in Kenya (Kitching, 2005).

As members of the worldwide construction community, we are able to bring protection and relief to society, by means of an effective and efficient construction industry.

Construction challenges

Global construction faces serious challenges, in spite of or due to its enormous magnitude. To accomplish its role effectively and efficiently, as expected by its numerous stakeholders, the industry needs to reduce, and preferably eliminate, a number of its shortcomings. These are not only limited to its dirty and old-fashioned image, with poor health and safety records and scant attention to environmental impact, as stated for instance by Sleight (2005b). Most construction markets also face increasing complexity, underdeveloped integrity and high failure costs. The latter are partly caused by out-of-date procurement practices, where selection procedures are still only based on the lowest price criterion. It is widely considered that the resulting number of disputes, claims and court cases is no longer acceptable, because of the resulting unreasonably high costs for both clients and contractors. In addition, ground conditions play a dominant role in the shortcomings. Examples of typical unsuccessful performance include excessive groundwater leakage in tunnels, slope instability, damage resulting from differential settlements and non-compliance with environmental requirements. In the words of Norbert Morgenstern (2000a): 'There are too many examples of unsuccessful performance in geotechnical engineering'. When combined, these factors put increasing pressure on construction quality, client satisfaction and profitability.

We will explore three of the main challenges for the construction industry, presented in Figure 2.1.

These challenges serve as a rationale for adopting risk management in our project practices and should motivate us to transform. The application of structured ground risk management may substantially alleviate them.

Increasing complexity

One of the transformation drivers in the construction industry is the increased complexity of many projects. One example is the design and construction of Terminal 2E of the Roissy-Charles de Gaulle Airport in Paris. Some 400 (!) different engineering and construction parties from all over Europe were involved in this 750 million euros project. Part of the terminal collapsed in May 2004, killing four people. How can we manage projects of this fragmented and complex character in a safe and cost-effective way?

Kevin Kelly is the author of *The New Biology of Business*. In his view, our technological systems have reached a complexity that acquires the characteristics of *biological systems* (Kelly, 1996). These can no longer be controlled by a mechanistic and linear method of management. It seems unavoidable that we will lose some of our control, while our conventional management of planning and control becomes obsolete. Terminal 2E may be an unfavourable indication of this. We must find

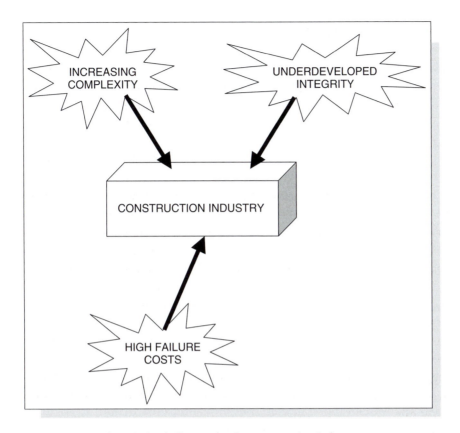

Figure 2.1: Three main challenges for the construction industry.

new ways to cope with this increase in complexity and its unavoidable partners: uncertainty and risk.

Because of the many stakeholders, all with their own and often conflicting interests, the construction industry is vulnerable to this increase in complexity. In addition, most construction projects are *prototypes*, as suggested by Wentink (2001). It is, in fact, what Kelly (1996) calls the *Hollywood-model*. Almost every construction project is unique, due to its project-specific purpose, stakeholders, expectations and of course the ground conditions and related challenges. In particular, medium-to large-scale projects operate temporary project organizations formed by consortia or joint ventures. These in turn hire subcontractors to provide specialist geotechnical and geoenvironmental services.

Finally, project complexity increases due to the demands of the external project environment and society. In densely populated urban areas, in particular, the surroundings are increasingly affected by and involved with construction projects. This leads to a new type of management, *stakeholder management*, as described for example by Edwards and Bowen (2005). It aims to manage and reduce the influence of external project stakeholders, such as environmental pressure groups. In addition, it strives to encourage these stakeholders to support project objectives, rather than obstructing them. There are numerous examples that could be given, but the message should be clear by now: today's construction activities are increasingly complex and this trend is not expected to diminish.

Underdeveloped integrity

A second challenge is to answer the growing aversion many people in a lot of countries feel towards *corruption* and *fraud* in the construction industry. This aversion has increased after numerous major business debacles in recent years, such as the Enron and Parmalat frauds and the substantially lower oil reserves than earlier presented by Shell. In 2002, the Dutch were shocked by what appeared to be the largest fraud case in their industrial history (see Box 2.2).

Box 2.2   Fraud and construction

The Dutch construction industry revealed its own integrity problems during a parliamentary enquiry in 2002–2003. Systematic problems included irregular pricing practices, artificial constraints on markets, as well as a degree of fraud (Process and System Innovation in Building and Construction, 2004). The latter, in fact, appeared to be a reaction to the introduction of new European regulations in 1992. Following this largest case of fraud in the Dutch industry, the Dutch construction industry is currently in a recovery and transformation phase. It has agreed to return some 70 million euros to Government clients. In 2005, a number of cases went to court. Twelve top executives of four major Dutch construction firms were prosecuted and found guilty of fraudulent activities.

The combination of fraud and construction is not exclusively a Dutch problem. Transparency International, a London based foundation, dedicated its 2005 Annual Report to the construction industry. It classified the construction industry as the most fraudulent industry worldwide. According to Transparency International, approximately 400 billion US dollars (equivalent to 400 000 million US dollars) is annually involved in corrupt construction throughout the world, with governments as clients. Public money that is not spent in the way it was intended. This figure represents some 10 per cent of worldwide expenditure on construction. In no other sector of the economy is the corruption percentage estimated to be so high (Transparency International, 2005). Box 2.3 illustrates that underdeveloped integrity is not limited to construction.

Box 2.3   Fraud and human behaviour

A small article in *NRC*, a high-quality Dutch newspaper, in September 2005, makes clear that the phenomenon of fraud is not at all limited to one particular industry. Recent research carried out by the Dutch association of insuring companies indicates that fraud using false insurance claims totals approximately 1000 million euros per year. This fraud figure is twice as high as previously assumed. Twenty per cent of all claims were investigated, of which 5 per cent turned out to be fraudulent. Is fraud an inherent part of human behaviour?

It is not just the ethical perspective that raises the issue of underdeveloped integrity. Corruption and fraud are also unfavourable from a business perspective. Without a level playing field with fair competition, honest and high-quality companies with interesting innovations are likely to be superseded by their fraudulent competitors. The result is less cost-effective projects and markets that do not serve society. Furthermore, the reputation of the industry becomes tarnished. Increasing numbers of bright and promising students are no longer choosing a construction career. We are losing these potential newcomers, who are so urgently needed in our complex industry.

Another unwanted effect of illegal practices is the impact on safety. Each year, the lives of tens of thousands of people are threatened by corrupt practices because safety standards are neglected. The previously-mentioned report from Transparency International presents numerous case studies in several countries, including but not limited to the Philippines, Uganda, Indonesia and Germany (Transparency International, 2005).

Underdeveloped integrity also slows down the economic development of people and regions. In the words of Heinz Brandl:

> The time is right for the engineering profession and construction industry to take a public stance against corruption. Corruption has – amongst other things – the effect of lessening the amount of capital invested in locations where infrastructure is often desperately needed (Brandl, 2004).

Organizations such as Transparency International create public awareness about the lack of integrity and its adverse consequences. Awareness is hopefully a first step towards long-term improvements. Explicit risk management, applied at every stage of a construction project, may further help identify corrupt practices. It requires a high degree of *transparency* in a project, an approach that conflicts with the inherent lack of clarity surrounding corrupt practices. Obviously, I am not under the illusion that risk management will act as an easy solution for the widespread and quite human phenomenon of corruption and its related practices. It may, however, at least help to create more project transparency for its stakeholders.

To conclude, we may join David Blockley and Patrick Godfrey (2000) in their book *Doing it Differently – Systems for Rethinking Construction*. This book won the Author of the Year Award 2000 of the Chartered Institute of Building (CIOB), the leading professional body for construction managers with over 40 000 members, working in over 94 countries. Blockley and Godfrey state:

> We believe that business ethics is an underdeveloped subject in the construction industry and the value systems that players use are not often explicitly discussed (Blockley and Godfrey, 2000).

High failure costs

Failure costs in the construction industry are sky-high. The third main challenge facing the construction industry is the need to reduce these high costs. Failure costs are defined as any costs resulting from unforeseen problems in construction projects. Typical causes are incorrect actions or deliveries, delay in the procurement of permits, (too) late design changes and particularly *unforeseen ground conditions*. These unforeseen problems often result from a lack of communication, information and time. They need to be solved, often during construction, which frequently puts severe pressure on budgets and planning.

Let us first consider The Netherlands. It is a rather small country with approximately 15 million inhabitants. Its construction market is significant, ranked as number 10 among the world's largest construction markets. Spending on construction totals some 70 000 million euros per year. Failure costs in the Dutch construction industry are estimated to be approximately 3500 to 9000 million euros annually. These failure figures are equivalent to between 5 and 13 per cent of the yearly expenditure within the Dutch construction industry. Table 2.1 presents a few 'high failure' Dutch infrastructure projects, together with their planning and cost overrun.

A typical example of excessive failure costs is the Tramtunnel Project. This cut-and-cover tunnel was constructed in the centre of The Hague, the political capital

Table 2.1: Planning and cost overrun of some Dutch infrastructure projects (after Molendijk and Aantjes, 2003)

Project	Overrun (%)	
	Planning	Costs
Dam: Ramspol inflatable dam	+80	unknown
Tunnel: The Hague Tramtunnel	+100	+90
Stormsurge barrier: Rotterdam Nieuwe Waterweg	+50	+40
Railroad: Betuweroute freight railway	unknown	+65
Railroad: Amsterdam to Utrecht	+65	+25

of The Netherlands. Construction costs for this project soared by 90 per cent, while the construction time increased by 100 per cent. Severe problems were caused by an innovative design, which did not give sufficient attention to groundwater leakage and associated fall-back scenarios. During construction, the building pit increasingly resembled a bathtub. Apart from the financial consequences for the parties involved, this project caused considerable inconvenience to the shopping public and shop owners in the city centre for several years.

Like underdeveloped integrity, high failure costs are not a typically Dutch phenomenon. Hoek and Palmieri (1998) evaluated a study of the World Bank, including 71 hydroelectric power projects. It indicated that costs and schedules were on average 27 per cent higher than originally estimated. For some of these projects, cost and time schedules escalated to several times the original estimates. In addition, some projects were stopped and abandoned entirely. Unforeseen geological conditions are not the main source for *all* of these cost and time overruns. However, the effects of inadequate geological data, their inappropriate interpretation and incompetence to deal with the resulting ground-related problems were significant.

There is abundant evidence that unexpected and unfavourable ground conditions are a dominant factor in many projects confronted with serious failure costs. Numerous case histories over the last century illustrate that failure to anticipate ground conditions is a main factor in construction problems (Fookes et al., 2000). Brandl (2004) indicates that, according to European statistics, about 80–85 per cent of all building failures and damages are related to problems in the ground. The Dutch Federation of Piling Contractors assesses their own failure costs at 100 million euros per year, equivalent to 10 per cent of their turnover.

Ground-related failure costs are, in particular, caused by unforeseen and unfavourable ground conditions. Contractual arrangements are often clear about which party is responsible for unforeseen ground conditions. Definitions of *what* ground conditions will be contractually considered as unforeseen are often remarkably vague, resulting in time-consuming and expensive disputes and claims, as described by Essex (1997). Gould (1995) presents the results of a study carried out by the United States National Committee on Tunnelling Technology in 1984. It relates the scale of site investigations to the occurrence of claims for 87 major underground projects. In total 60 per cent of the 87 projects were confronted with claims, of which 95 per cent concerned substantial amounts. While 64 per cent of the claimed amounts were awarded, overall construction costs increased by 12 per cent, not including additional legal costs.

Acceptance by many construction stakeholders of time and cost overruns is decreasing dramatically. Many members of the public, for instance, find it difficult to accept that the first people were walking on the moon in the 1960s, while today's construction on our planet earth still cause a range of serious problems. For clients and owners of construction projects, failure costs have adverse effects on their budgets, planning and reputation. For engineers and contractors, failure costs have a serious impact on their planning, reputation, as well as on their profitability. In the Dutch construction industry the rather low profit margins of 2–4 per cent are under attack by ever increasing competition (van Staveren, 2001).

In conclusion, failure costs result in a waste of money, time and energy for most of the stakeholders involved. Rather than bringing projects to a minimum acceptable level of quality, it is far better to use these resources to reduce costs further, or to increase profits and quality. I am the first to agree with those readers who will argue that we will never be able completely to eliminate failure costs in construction. However, we can undertake serious attempts to reduce them, particularly with regard to ground conditions.

Crisis – what crisis?

Some readers, with pessimistic or even realistic viewpoints, may classify the state of the construction industry as one of *crisis*. The previously presented figures and examples highlight the need for transformation. Improvement of control over ground conditions emerges as one of the key success factors for construction. Since such control often remains underdeveloped or clouded by activities of a doubtful integrity, some of us may indeed conclude that there is a silent and hidden ground-driven crisis. The need to give transparent and adequate attention, in a timely way, to the inherently uncertain and risky ground, appears more than ever necessary for a prosperous construction industry. This may require an equally dramatic change of perception, as indicated by Box 2.4.

Box 2.4   Crisis and perception

The physicist Fritjof Capra published his book *The Turning Point* in 1983. In many countries, the 1980s were considered as a period of crises. Many societies struggled with high inflation, unemployment, environmental problems caused by pollution and a rising wave of violence and crime. Little appears to have changed since then. According to Capra (1983), *perception* is the reason for the crises. Outdated worldviews need to be replaced by new ones. As a physicist, he recalls the crisis in physics in the 1920s, when the mechanistic worldview of Newton became replaced by, or at least accompanied by, the view of quantum physics, with its dynamic and at times even chaotic character. Perhaps we should attempt to change our perceptions of construction?

We have so far explored the ever-increasing complexity, the case for integrity and the high failure costs associated with today's construction industry. However, there is no reason to accept this situation. Mankind is able to initiate and complete amazing achievements. Why then, should we not transform our construction industry towards a prosperous one? An industry which can deal effectively with complexity, where integrity is developed to acceptable standards, and where failures costs are reduced to acceptable levels, in order to allow reasonable profits? The opportunities are present and initiatives for change have already started in a number of countries!

Opportunities for the construction industry

Construction needs a renewed *licence to operate*. As many books about change management explain, any change must start with *awareness* for its need, followed by the willingness to act accordingly. The preceding sections in this book are intended to serve as a loud wake-up call. The following sections explore opportunities to transform our industry, varying from a concise presentation of some of the ongoing change initiatives, to the role of systems thinking and risk management. Finally, attention is given to the people factor, in accordance with the high-tech and high-touch approach of this book.

World-wide change initiatives

The construction world is ready for change. In a lot of countries, a variety of stakeholders asked themselves how the construction industry could become modern, competitive and innovative. These questions have resulted in industry-wide initiatives, often supported by government. In Europe, change initiatives have taken place in Denmark, Finland, Norway, the UK and The Netherlands. Outside Europe, similar initiatives are running in Hong Kong, Singapore and Australia. In 2004, a Dutch delegation representing one of the main Dutch change initiatives, Process and System Innovation in Building and Construction, visited these countries to exchange experiences and knowledge (Process and System Innovation in Building and Construction, 2004). I present their most remarkable insights below.

The driving force for change in *Norway* is the offshore industry coming to land. Close relationships with the offshore industry support the development and implementation of quality management, cost engineering and standardization. A Life Cycle Costing (LCC) approach is mandatory for all public buildings in Norway. It enhances the awareness of total costs of ownership and clients' needs. In 2003, the main players in the Norwegian construction industry agreed a common code of ethics.

Corruption scandals in the 1980s and 1990s triggered *Hong Kong's* reform initiatives in the construction industry. Serious problems included the need to demolish high-rise residential building blocks, for safety reasons, because of insufficient piling. In 2000, Hong Kong set up a Construction Industry Review Committee (CIRC) to manage the construction industry transformation processes.

Singapore started the Construction Manpower 21 Study in 1998. It aims to transform construction from the three Ds – dirty, demanding and dangerous – towards the three Ps – professional, productive and progressive – to create a professional and knowledge-based industry.

Labour disputes, contractual disputes and some fraud cases drove the reform process in Australia. A number of initiatives were initiated in the 1990s, at both commonwealth and state levels. The Australian experience demonstrates that time is required to create real change. A clear purpose and steady commitment have already shown, however, that it is possible to change a culture of suspicion, corruption and lack of trust into one of collaboration and mutual respect.

The Egan Report, titled *Rethinking Construction* and published in 1998, challenged the construction industry in the *UK* (Construction Task Force, 1998). The Egan challenge is ambitious, with targets including an annual reduction of 10 per cent in construction costs and time, as well as a 10 per cent increase in turnover and profit. Despite the fact that Blockley and Godfrey (2000) recognized a degree of initiative fatigue due to the many initiatives in recent years, some have been shown to work. Pilot projects, coordinated by Construction Excellence, provide some impressive figures, such as a 2.7 per cent increase in safety and a 3.5 per cent fall in personnel turnover. One pilot project even showed a 65 per cent rise in productivity, in terms of added value per worker. Furthermore, construction costs for these pilot projects were on average 6 per cent lower and profitability increased 2 percentage points (Heijbrock, 2005).

Three Dutch ministries jointly established the Dutch Construction Steering Committee in *The Netherlands*, in 2004. The committee's mission is to transform the Dutch construction sector within the next four years. Three themes were selected:

1   Innovation

2   Transparency

3   Quality-price relationship.

The third theme should provide an escape route from the lowest price trap, by applying innovative procurement and contracting methods. These themes respond to the three main challenges presented earlier. In their first annual report (Dutch Construction Steering Committee, 2005), the committee recommends a move from a mainly risk-adverse attitude towards a more risk-seeking one. Cooperation between different parties in the construction process should result in *risk sharing* instead of only *risk allocation*. It seems that risk management has gained a stake in the Dutch transformation process. Its concept and tools will be used as a communication vehicle to transform the industry from being rather risk-unaware and closed towards a more risk-aware and open industry. The concept of systems thinking may act as catalyst.

Systems thinking

Peter Senge introduced the concept of *systems thinking* into management in the 1990s, in his best seller *The Fifth Discipline – The Art and Practice of the Learning Organisation*. Systems thinking *is* the fifth discipline. It integrates the four other disciplines of personal mastery, awareness of mental models, the capability to build a shared vision and team learning (Senge, 1990), as illustrated in Figure 2.2.

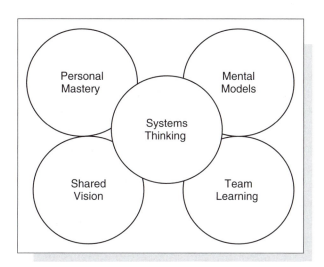

These five learning disciplines are *personal* disciplines, distinguishing them from the more familiar and collective management disciplines. These personal disciplines are essential for creating a learning organization, where professionals can discover how to create their own reality and how to change it. In Peter Senge's own words:

> At the heart of the learning organisation is a shift of mind – from seeing ourselves as separate from the world to connected to the world, from seeing problems as caused by someone 'out there' to seeing how our own actions

Figure 2.2: Senge's five learning principles (after Senge, 1990).

create the problems we experience (Senge, 1990).

This view needs to be adopted by the construction industry, which is also the opinion of Blockley and Godfrey (2000). They present an extensive set of theories, tools and practices around the concept of systems thinking, largely building on the work of Senge and translating it to the construction industry. One of their main ideas is *thinking differently*. In their view, construction should be considered as a *construction process*, with substantial human input and consequences at every process stage, rather than just creating a *construction product*. Applying the systems thinking concept facilitates this process approach.

Blockley and Godfrey (2000) introduce two fundamentally different types of system that are also closely interrelated:

1 Soft systems

2 Hard systems.

A *soft system* is characterized by its human and subjective elements. It includes the people factor and is comparable with high-touch (Naisbitt, 1984). Soft systems are related to social sciences, management and marketing. They not only comprise the well-known triggers of action and reaction, but also their *intention*. A soft system also incorporates *reasoning* about *why* we do or why we do not. For instance, the appraisal and improvement of team-performance and the exploration of a client's needs are typical aspects that require a soft systems approach. The same applies to decisions about the type and extent of site investigations, because this is ultimately driven by the client's highly subjective risk tolerance.

Many technically-educated readers are likely to be more familiar with the second type: the *hard system*. Contrary to a soft system, a hard system is physical and objective. It excludes the fuzzy human factor and is comparable with the high-tech approach (Naisbitt, 1984). The discipline of physics, which serves as foundation for the applied engineering sciences, is typically a hard system. Geotechnical design calculations are also hard systems, if we exclude the rather subjective character of their input parameters. Pure hard systems comprise only action and reaction and exclude vague human elements such as intention.

Hard systems are *always* embedded in soft systems. It is ultimately mankind who develops and operates all system types, including the hard ones. There are ample reasons for focusing on this soft system–hard system interrelationship. Three of them have already been explored: increased complexity, underdeveloped integrity and high failure costs. Their associated problems cannot be solved by a hard system approach alone. Unfortunately, working life is not that simple. We should consider the whole system, the hard high-tech cores and the softer high-touch shells around them, in order to transform the construction industry in a structured and irreversible way. The entire construction processes need to be considered, from the earliest beginning through to operation and maintenance.

This is easier said than done. Communication between hard and soft systems is inherently difficult, because of their fundamentally different characters. As many of us will recognize, it is the same type of friction between the sales department and the construction department in many companies. Representatives of both departments communicate from a different viewpoint, from a mainly soft and hard system approach respectively. Which is where *risk management* may help.

From systems thinking to risk management

Risk management can be embedded in the *twilight zone* between hard and soft systems (Figure 2.3). The zone boundaries are deliberately not shown as a solid line to demonstrate that soft and hard systems are not closed in reality, as we often like to assume for reasons of simplicity.

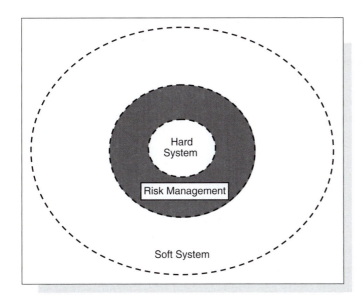

Figure 2.3: Risk management in the twilight zone of a hard and soft system.

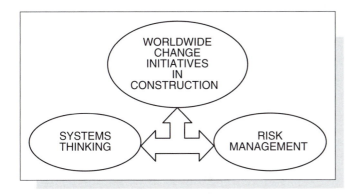

Figure 2.4: The relationship between change initiatives, systems thinking and risk management.

Risk management can be seen as a communication vehicle between soft and hard systems. It can act as the translator between both, as it speaks two languages. The *technical* or hard systems language is the mathematical one of statistics and finite element analysis. Quantitative risk analysis is a typical hard systems exponent. Its *human* or soft language equivalent is that of dealing with the inherently different risk perceptions of professionals and other involved parties. It includes the wide area of qualitative methods for identifying and classifying risks.

Systems thinking and risk management complement each other and can be expected to facilitate the worldwide change initiatives in construction (Figure 2.4).

The combination is particularly promising for ground-related risk management, as most ground-related problems include subjective and soft elements, in addition to the conventional objective and hard elements. The GeoQ concept is a mixture of hard and soft system components. Regarding the soft component, any risk-driven action should be motivated by the risk perception of those stakeholders responsible for that risk. Hard systems or high-tech deterministic models are required as well. They provide data and arguments for the physical aspects of the project. However, these hard systems always remain embedded in their risk-driven and human context.

In view of all the challenges and opportunities described for the construction industry, the ability to appraise and implement risk management will undoubtedly become a critical success factor for many firms and organizations if they are to survive in today's construction arena. In this respect, it may be helpful to remember the following words of Peter Senge:

> Systems thinking shows us that there is no outside; that you and the cause of your problems are part of a single system. The cure lies in your relationship with your 'enemy' (Senge, 1990).

We may need some renewed inspiration, to keep these words alive in day-to-day practices.

Renewed inspiration

Which of us would not like to help to transform our industry, in order to realize some of the promised cost savings, higher profits and other benefits? The application of systems engineering and (ground) risk management will certainly help us, but will not be sufficient in isolation.

   Why is it that initiating change is always so difficult in practice? Why is it that many of us acknowledge the need for rigorous and irreversible reform, but still struggle, day after day, with fierce competition using only the lowest price criterion? Libraries are filled with books about *change management*, but we still do not have the definite answer. Obviously, I do not have *the* answer, either. But I do have a clue about *an* answer, based on inspiration and perception, as presented by the tale in Box 2.5.

Box 2.5   Inspiration and perception

More than 4500 years ago, two workers were cutting stones in Gizeh, Egypt. A woman passed by and asked them what they were doing. The first worker answered: 'I am cutting stones, as you could see'. The second worker answered: 'I am contributing to one of the most famous construction projects ever'. Both persons were doing exactly the same work.

   I recognize an increasing call for inspiration among many professionals in a variety of roles and functions. There seems to be an underestimated need, which is not limited to the construction industry. Young professionals, in particular, a new generation embarking on their working lives, are anxious to make a difference. We do not need to build pyramids to be inspired. The impacts and challenges posed by our own construction projects should be more than sufficient.

One fundamental requirement is needed to channel our inspiration. I refer to it as individual empowerment and it is closely related to the phrase of Peter Senge at the end of the previous section. We need to be firmly dedicated to our own contribution, by following our own unique capabilities and characters. It is very much an *individual choice*. Changing organizational structures and culture is often seen as the way to implement change in an organization, whatever these concepts may be. These changes will not result in real and long-term change, however, without the dedicated willingness of individuals, such as you and me, to change (van Staveren, 2005). This places a great deal of responsibility on our shoulders.

Summary

The magnitude of construction is enormous, as demonstrated by the figures and statements included in this chapter. A number of major challenges face the world's construction industry, which therefore influence our daily lives as well. We have recognized ever-increasing *complexity*, an underdeveloped sense of *integrity* and substantial *failure costs*. In all of these, *ground conditions* play a major role.

Is there an ongoing crisis in our industry? Perhaps there is, but the good news is that new solutions and opportunities arise. In recent years, a number of countries have initiated ambitious change programmes for the construction industry. They share one main objective: to transform it into an industry where sustainability, safety, integrity, and profitability are increased. We have considered some of these initiatives, and they may help us cope effectively with our challenges.

The concept of *systems thinking* is seen as one of the keys to unlock possibilities for the industry transformations required. It stipulates fertile ground for the concept of *risk management*, which may act as a link between the hard and soft systems of construction.

We should realize that only *people* can bring about any industry transformation, by embedding ground risk management in daily practice for instance. These professionals can make it happen, but they may need renewed *inspiration*. This chapter has prepared the way to explore a number of concepts about uncertainty in the next chapter, as well as risk and risk management in relation to ground conditions.

3 From uncertainty, risk and ground to GeoQ

Introduction

This chapter is about a number of *concepts* or mental models of simplifying the world (de Bono, 1998). Concepts are, by definition, a little bit fuzzy and general, but they prevent us from getting stuck in details in an early stage of thinking. These characteristics make concepts so useful and I will use them a lot in this book and particularly in this chapter.

This chapter starts with the concept of *uncertainty*, followed by the concepts of *risk* and *risk management*. Many books are dedicated to these topics and I have selected those aspects that are expected to support us in our practice of ground risk management. Next, the concept of *ground conditions* is related to the concepts of uncertainty, risk and risk management. Ground risk appears to be of a rather hybrid nature. An overview of the main types of ground and ground-related risk serves as a basis for the introduction of the GeoQ ground risk management concept.

The introduction of the GeoQ concept starts with some remarks on the current status of ground risk management worldwide and a brief history of its recent development. Then the GeoQ concept presents itself, including its perceived benefits. It is related to the existing ground-related disciplines and positioned within the risk and hazard management landscapes. As the letter Q in GeoQ is derived from the word quality, also the role of GeoQ with regard to quality management gets attention. The last section of the chapter positions GeoQ towards knowledge management and the chapter ends with a summary.

The concept of uncertainty

Uncertainty is a certainty. We all encounter uncertainty, in our daily professional and private lives. *Uncertainty* can be defined as an absence of information about parts of a system under consideration. Uncertainty is always present, even when information is perceived as complete (Smallman, 2000). Of course, there are some certainties in this world, such as we will all eventually die. And if I jump from a crane, I will certainly fall, according to Newton's well-known law of gravity. However, even in the apparent objective world of physics, there seems to be reasonable uncertainty, as explained by a historical example in Box 3.1.

Box 3.1   Even sound physics is highly uncertain

In 1927, the physicist Werner Heisenberg published his so-called *uncertainty principle*, which was a real break-through in the development of the, at that time, new discipline of quantum physics. He stated that if you consider a very small particle, the so-called quantum particle, there is only a *probability* about the location of that particle in its quantum space. His colleague Niels Bohr increased the uncertainty in the same year, by stating that the same quantum particle can act both as a particle and as a wave, his so-called *complementarity principle*. These principles upset the famous Albert Einstein, who wanted to see an objective physical reality. He made this clear by speaking his famous 'God does not play dice' (Ortoli and Pharabod, 1986).

I make this side step because physics is commonly considered as a rather objective discipline, with a high degree of certainty and repeatability. Quantum physics taught us that our common and successfully applied laws of mechanics have their limits when we enter the microcosmos of our material world. Similarly, a lot of technological solutions approach their limits as well, for instance when we enter the macrocosmos of our construction projects in interaction with the public. Later, we will explore this issue in more detail.

Three types of uncertainty

According to Blockley and Godfrey (2000), we have to distinguish between three types of uncertainty: *randomness, fuzziness* and *incompleteness*. We will consider each of these from a ground conditions perspective. They are presented in arbitrary order of importance towards ground engineering. I use the definitions of Blockley and Godfrey and relate them to some common aspects of ground conditions.

Randomness
Randomness can be defined as a *lack of a specific pattern* of individual parameters or variables. However, there may be patterns identified from populations of data. In other words, if we consider a set of parameters, such as the unconfined compressive strength of rock or the undrained shear strength of soil, then a pattern may

become visible. These patterns can be presented by probability-density functions. They define the distribution of the probability of occurrence of these ground parameters. A well-known pattern in statistics is the bell-shaped normal distribution, which is determined by the mean value and standard deviation of a particular parameter (Anderson et al., 1999). These statistically derived patterns may give at least some apparent certainty. We may think that we can make predictions based on these statistics. We may derive parameters, like the mean value and standard deviation. Modern software makes it easy to present the results in nice and colourful pictures and diagrams and allows us to create all kinds of apparent relationships, with an apparent sound statistical basis.

However, we must realize that these types of relationships between variables are not necessarily the result of cause and effect. It is a fact that is easily forgotten in today's digital world, at least with some less experienced professionals. For instance, sound theories about the geological history are still required to understand and rely upon statistically derived relations between ground parameters. This knowledge can be of great help to get more insight into certain patterns of ground characteristics. A well-known example is the grain size distribution of sand and gravel, which can be related to the deposition of that material by meandering rivers (Press and Siever, 1982). Unfortunately, apart from a lot of geological theories for scientific purposes, I am still missing a lot of accessible and practical tools to account for geological heterogeneity in ground engineering and construction. Also, in rock engineering, we encounter a lot of random uncertainty, for instance with regard to weathered rock and tropical residual soils. While considering these restrictions of ground-related statistics, its entrance within ground engineering is a fact. If used with care, it will provide a lot of additional and useful information, as further explored in Chapter 10.

To conclude, it can be very helpful to be aware of the likely randomness of ground as a source of uncertainty, especially in those situations where we do not have the geological evidence for the existence of clear patterns. This is often the case in geologically complex areas, with a rich history of different types of subsequent deposition, erosion, glaciation, and so on. In ground-related disciplines there is often a lack of sufficient data, which are necessary to disclose any patterns with an acceptable statistical certainty. Therefore, often these patterns, if at all present, stay hidden in the ground. Consequently, in many construction projects many ground related data have to be considered as random, which creates, for example, uncertainty about what parameters to use as input in calculation models for geotechnical design.

Fuzziness

Fuzziness is the second type of uncertainty as distinguished by Blockley and Godfrey (2000). In relation to ground engineering, it is also important to be aware of this type of uncertainty. Fuzziness can be defined as *imprecision of definition* or *imprecision of concept*. Our highly valued concern of precision is challenged by fuzziness. As a matter of fact, we encounter the limitation of language. We approach the restrictions of objective communication between people, as is illustrated by the philosophical struggle in Box 3.2.

Box 3.2 Fuzziness and the problem of communication

The French philosopher and writer Jacques Derrida (1930–2004) is considered to be the founding father of the concept of *deconstruction*. It is a critical analysis of that which seems apparent and clear in our language and writing. I relate this deconstruction to the concept of fuzziness in our communication. In fact there is an existential subjectivity because of the limitations of the objective logic of our language and writing. My intention is not exactly the same as what I think. What I think is not exactly the same as what I write or say, which is not equal to what you read or hear. Finally, what you read or hear is not similar to what you interpret and understand. Our own unique perceptions and interpretations of the world around us cause it. In addition, our language, even supported by rules of grammar and dictionaries, is inappropriate to express these subtle subjective individual aspects. According to Derrida, objective truth is non-existent. True or not, that is a philosophical question. At least we should consider our communication as a rather fuzzy process, which is an additional obstacle to deal with uncertainty.

Obviously, with regard to Derrida's non-existent objective truth we should make exceptions, such as for the mechanical laws of nature, so long as we do not enter the microcosmos. The traditional mechanical laws prove to work well in our normal day-to-day conditions, as well as its objective language of *mathematics*. However, this language is only applicable in what we earlier defined as the hard systems. It is difficult, if not impossible, to apply mathematical expressions for soft systems with subjective perceptions, such as importance, priority and beauty.

What does this concept of fuzziness imply in our ground engineering and construction practice? To start rather simply, for instance, the statement of *soft rock* is fuzzy, because of the imprecision of the word soft. What is meant by soft? How soft is soft? In what range of kPa can the unconfined compressive strength of the soft rock be expected? Within ground engineering this type of fuzziness can be easily countered by the use of one of the many available classification systems. Many countries have their own classification systems for rock, soil and intermediate materials, which are adapted to the local geological conditions. Other classification systems are of a general character and can be applied in many parts of the world. Examples are the British Standards, the Unified Soil Classification System and the American Standards for Testing and Materials. Therefore, this simple type of fuzziness is probably not a major problem within the ground-related disciplines. Ground engineers are used to coping with this type of fuzziness.

Box 3.3   Fuzziness of ground sampling

The following three examples consider *ground sampling* in a fuzzy perspective:

1   *Ground sampling* always results in some form of sample disturbance. So-called undisturbed samples, often mentioned in the literature, simply do not exist. At best, ground samples with a relatively low disturbance can be retrieved by special sampling techniques. The type of ground itself is also a dominant factor in the degree of sample disturbance. Highly weathered rock and very soft soil are very difficult to sample in a good enough way. But how do we *measure* sample disturbance and consider its effects on the outcome of laboratory test results? How do we allow for these effects in the selection of design parameters? How do we account for the effect of sample disturbance in geotechnical design?

2   *Cone Penetration Testing* (CPT) is widely used in many parts of the world to obtain rapid and high quality information about the type and strength of soil and highly weathered rock at reasonable costs. However, the ground collapses while the cone of the CPT measures ground resistance during penetration. It is, in fact, a destructive type of research, so what is the real strength of the ground, just before it collapses? What is the effect of the design of the cone? How do we deal with the measured friction resistance between the ground and the cone, a typical result of fuzzy interaction? In spite of its high scatter, friction resistance is widely used for classification purposes, which makes it quite unreliable for classification without benchmarking by borings (van Staveren and van Deen, 1998). The work of Lunne et al. (1997) provides an extensive overview of all these fuzzy factors.

3   *Environmental sampling*, in which samples of polluted groundwater or polluted ground are taken, is very sensitive to so-called *cross-contamination*. Clean soil layers can become polluted during ground or groundwater sampling. There are numerous cases in which this type of cross-contamination occurred, making the problems even worse than before sampling.

However, sometimes professionals are unaware of the fundamental implications of ground fuzziness, such as increasing ground risk or neglecting hidden opportunities. If we go a step deeper into ground engineering and construction, we will discover a lot of fuzzy uncertainty in all kind of ground-related issues. The following six examples about fuzziness and ground sampling (see Box 3.3) and ground engineering (see Box 3.4) may help to reveal ground fuzziness in our practice.

The examples of fuzziness in Box 3.3 are a result of the *interaction* of the ground sampling or testing technique with the ground itself. It provides an interesting similarity with experiments in the human sciences, like psychology and sociology, where there is also some kind of interaction between the person doing the test and the tested person. Apparently, ground sciences and human sciences are not that different.

Also in the field of *ground engineering* we can see a lot of fuzzy uncertainty, as presented by three examples in Box 3.4.

Box 3.4   Fuzziness in ground engineering

The following three examples consider *ground engineering* in a fuzzy perspective:

1   The calculation of *horizontal deformation of foundation piles*, for example, caused by horizontal loads of an embankment, needs a lot of complex interpolation steps. The *interaction* of the ground and pile deformation necessitates these interpolations.

2   What type of *elasticity constant, E-value*, should we select in order to calculate the horizontal deformations of the previous pile example? Recently, I was supervising an MSc student researching this type of problem. She identified some 30 different E-values. What type of E-value should be selected for what type of calculation?

3   The calculation of the *bearing capacity* of hollow steel piles, as in the offshore construction of oil production platforms and jetties, includes a considerable degree of fuzzy uncertainty as well. Is it permitted to include the bearing capacity of the ground inside the hollow pile in the total bearing capacity? That ground is disturbed during the installation of the pile. Neglecting the bearing capacity of the soil plug within the pile would be safe, but may result into an expensive over-design of the foundation.

In conclusion, there is a lot of fuzzy uncertainty in ground sampling and ground engineering. I am sure that many of the readers who are directly involved in design and construction with ground can add their own examples. Of course, experienced engineers have found practical and technically acceptable solutions to deal with this fuzziness, often by modelling and simplifying reality. In practice, we can deal with it in one way or another. However, there is a *risk* that we explicitly do not acknowledge this fuzziness anymore, because we are so used to handling it. We deal with this fuzziness in an implicit way. If we want to bring the ground-related engineering and construction in a risk driven context to create more economic structures, with at least the same safety, then *revisiting fuzziness* is an ultimate requirement.

Incompleteness

Finally, we arrive at the third type of uncertainty described by Blockley and Godfrey (2000), which is *incompleteness*. With this type of uncertainty we will feel really at home in our ground-related disciplines because of the inherent lack of complete data from the ground. *Incompleteness* can be defined as *missing information*, that which we do not know. Two subdivisions of incompleteness can be made in this respect. The first one is *that which we know that we do not know*. This type of incompleteness is common within ground conditions. Between two boreholes for instance, we know that there is some kind of ground, of which we do not know the precise characteristics. This type of uncertainty is *foreseeable*. As we will see later in this chapter, the types of risk that may occur from this type of uncertainty can be well managed.

The second type of incompleteness has a more difficult character. It concerns *that which we do not know that we do not know*, if you still can follow me. In other words, it is about a total unawareness of what we do not know. To go back to the example of the two boreholes, it is about an *unforeseeable* uncertainty of the material between the two boreholes. It will be clear that management of the unforeseeable is, by definition, impossible.

In our projects we can minimize the 'unforeseeable' by considering as many different views and opinions from as many different experts as economically viable. To wrap up about the uncertainty type of *incompleteness*, it is a given when dealing with any type of ground condition in our projects. Incompleteness is at least one of the few certainties we have!

Before ending this section about uncertainty, there is an additional remark to make. The *Bayesian Belief Theory* involves both the *randomness* type of uncertainty and the *incompleteness* type of uncertainty. You may encounter it in the literature and therefore I introduce it briefly in Box 3.5.

What are the possible impacts of these ground uncertainties on our construction projects? To answer this question, we need to explore the concepts of risk and risk management.

Box 3.5 Combining random and incomplete uncertainty

A well-known method for dealing with the interpretation of so-called random variables is the *Bayesian Theory* (Bles, 2003). In this theory, probability is not *measured*, such as rolling the well-known dice, but is based on the experience of experts. With its inherent subjectivity, the latter can be called the professional *belief* of these experts. These beliefs are clearly not only based on hard factual measurements, a reason why Bayesian Theory is sometimes referred to as *Bayesian Belief Theory*.

A remarkable element of this theory is that it does not allow incompleteness, because all probabilities must sum up to one. In other words, it needs complete information. Indeed, the absolute given factor in all ground-related aspects, incompleteness, is necessarily excluded in this Bayesian Theory. However, Bayesian Theory has been entering geotechnics, despite the fundamental limitation of required completeness of information. While taking explicit *assumptions* to arrive at complete information, the Bayesian approach can give useful additional information (Bles, 2003).

The concept of risk

Some definitions

No risk – no glory. The term *risk*, and the associated term *hazard*, have a lot of definitions, proposed by numerous experts and institutions. *Hazards* can be defined as threats to people and the things that they value. In the view of Carlsson et al. (2005), an uncertainty becomes a risk if a probability is assigned to it. *Risk* can be defined as the product of the probability and the eventual impact of a hazard (Smallman, 2000). In other words, risk is the product of the probability or likelihood of an undesired event and the consequences of that event.

Indeed, it means that the uncertainty of the risk event is twofold. First, the uncertainty about the occurrence of the risk, which can be considered as the probability part. Second, when the event occurs, the uncertainty about the likely consequences, which can also be expressed in terms of probability. In the context of risk management, uncertain events are usually considered as hazards with the potential to have *negative* effects. However, uncertain events can occasionally also give attractive opportunities with *positive* effects, as is reflected in a perception of risk in Box 3.6. Risk may even create a new emerging market.

Box 3.6   Risk is the gold of the 21st century

Saskia Sassen, professor at the University of Chicago and the London School
of Economics, made this statement during the third Ernst Heijmans Lecture in
November 2004, Utrecht, The Netherlands. Because of ongoing globalization,
the complexity of a lot of situations and cases for a majority of firms increases.
This is, in her opinion, the cause of the emergence of very specialized service providers and consultants in accounting, legal affairs, design and so on.
These services are provided in no matter what market sector, thus including
the construction sector. Sassen acknowledges the emergence of what she calls
an *intermediate economy*. This type of economy will fill an increasingly strategic
space between, in her words 'the monsters of uncertainty and their clients'.
These clients are firms who choose to outsource their risks to this intermediate economy (Sassen, 2004). In this view, uncertainty and risk are developing
towards products of trade, perhaps even towards the gold of the 21st century.
What are the effects of this risk-as-trade movement on ground-related risk in
the engineering and construction industry?

I have a clear rationale for the above proposed simple definitions of the terms
risk and hazard, because I am a big supporter of the so-called *KISS principle* –
Keep It Simple and Short. The presented risk definition proved its practical value
in ground-related engineering and construction. It is left to the reader to use
more complex terminology, when appropriate and viable. Normally, that will
only be the case in the later and detailed stages of risk management, when it has
become clear that particular identified risks need a careful decomposition, in order
to manage them. Based on my experience, it is a common pitfall to make risk
management too complicated in the early stages of any project, which confuses
and frustrates the people involved. Often, this results in the unfavourable end of
any structured risk management.

Risk types

Many different *risk types* can be distinguished. Three main and practical types
of risk will be presented below, *pure* and *speculative risk*, *foreseen* and *unforeseen
risk* and finally *information* and *interpretation risk*, because these types facilitate the
application of the GeoQ process.

Pure and speculative risk
The first distinction is between *pure* and *speculative risk*. *Pure risks* are, by definition,
related to hazards or unwanted events and, therefore, always have an undesirable

outcome. For these risk types, successful risk management can never be better than the entire elimination of the hazard. *Speculative risks* can have both desirable and undesirable outcomes. Successful risk management for speculative risks means maximizing its opportunities. Edwards and Bowen (2005) warn that catching opportunities should not replace remediation of those risks that threaten the project objectives. Pure and speculative risks interact, therefore, both sets of risk should be considered, ideally in a holistic way (Waring and Glendon, 1998). Box 3.7 illustrates how the distinction in pure and speculative risk coincides within the Chinese language.

Box 3.7   The Chinese approach to pure and speculative risk

Since the new millennium the Chinese economy has been booming. Business is often related to taking risks. Therefore, we may ask ourselves, is there something special about the risk attitude in China? Considering the Chinese language, this may be the case. The speculative and potentially positive effects of risk appear to be reflected in the Chinese language, because the words *crisis* and *opportunity* have the same Chinese character. In other words, in the Chinese perception, risk resulting in crisis is equal to opportunity. Apparently, risks and opportunities are perceived as two sides of the same coin. What can we learn from this attitude?

This opportunity driven approach to risk is embedded in this entire book, because I am convinced that such a positive approach will finally help to establish risk management in ground-related engineering construction activities. Figure 3.1 relates the three types of uncertainty, as introduced in the previous section, via opportunities and threats, to speculative and pure risks.

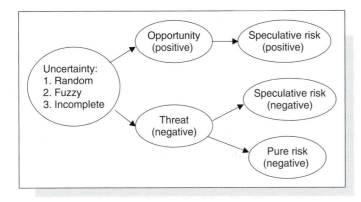

Figure 3.1: A simple relationship between uncertainty, pure risk and speculative risk.

As Figure 3.1 shows, all three types of uncertainty can be considered as both opportunities and threats. Opportunities relate only to speculative risk, while threats represent either the negative type of speculative risk or the pure risk.

Foreseen and unforeseen risk

When considering ground risk management, another useful distinction is between *foreseen* and *unforeseen* risk, as for example recognized by Altabba et al., (2004). Not surprisingly, *foreseen risks* are foreseen during the risk identification process. They are the main subject of risk management and some examples are differing site conditions, adverse weather conditions and failing con-

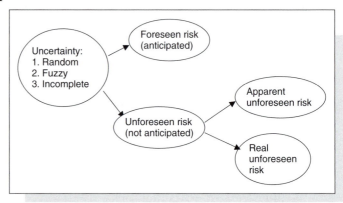

Figure 3.2: A simple relationship between uncertainty, foreseen risk and unforeseen risk.

struction equipment. More of a problem is the *unforeseen risk*, caused by negative and unwanted events that are *not* anticipated during the risk identification process, such as risk of political origin and risk that is caused by unexpected market fluctuations, which can result in severe price effects. In fact, even these risks are not really unforeseen, as we recognize them already in some way. It is better to consider these as *apparent* unforeseen risks, with both highly uncertain probabilities and consequences. *Real* unforeseen risks remain unknown to us, by definition. Foreseen and unforeseen risk types may originate from all of the three types of identified uncertainties, as shown in Figure 3.2.

Both apparent and real unforeseen risks are typically located outside the circle of influence of all project stakeholders, such as the project teams of the engineer and the contractor, the client and the end-users of the project, which does not make them popular.

Information and interpretation risks

Particularly in relation to contractual matters, it is common and helpful to distinguish between *information* risks and *interpretation* risks. Information risks are about wrong and incomplete factual data. It is an *objective* risk and includes the uncertainties of *randomness* and *incompleteness*, as previously described in the approach of Blockley and Godfrey (2000).

Interpretation risk is about different and inherently *subjective* ways of interpreting factual ground data. Differences in interpretation may arise from a wide range of sources. Examples are different views on the interpretation of geological hetero-geneity, incorporation of the effects of ground sample disturbance in the selection of design parameters and decisions between analytical or finite element calculation models. Interpretation risk aligns with the fuzzy type of uncertainty. Its reduction is consequently complicated, due to a wide range of sources of fuzzi-ness. Figure 3.3 presents the relationship between the three types of uncertainty, information risk and interpretation risk.

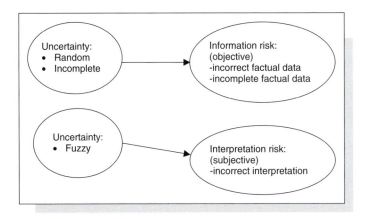

Figure 3.3: A simple relationship between uncertainty, information risk and interpretation risk.

It is remarkable that *incorrectness*, as part of the information risk, is lacking in the three types of uncertainties. Because ground engineering and constructions are performed by human beings, the uncertainty and associated risk of incorrectness, in other words, human errors, must not be neglected. As stated by Cummings and Kenton (2004): 'The fundamental (real) cause of failures is human failure'. They admit immediately that most projects do not fail. I agree with them, while acknowledging the relatively high failure *costs* in construction.

Dynamic character of risk

Another and elementary aspect of risk is its inherent *dynamic* character. One simple but convincing factor is the time-dependent consolidation behaviour of cohesive soils like clay. The geologically determined history of a clay layer affects its compressibility and consequently the risk of settlement. Pre-loading by ice, during periods of glaciation, or simply by an ancient construction, will markedly change the clay strength and deformation properties.

However, often, probably for reasons of ease, risks are considered as *static*. This assumption makes risks much simpler to manage. Unfortunately, reality is complex and risk management is too. About 3000 years ago, the Greek philosopher, Heraclitus from Ephesus, spoke the wise words *pantha rhei* – everything flows (Aufenhanger, 1985). Look carefully around you for a moment, wherever you are. Can you mention something that is *not* subject to change, on the short, the longer or even the very long term? Even we are changing, now at this very moment. Millions of cells of our own body are renewed every day. Furthermore, we will grow older, whether we like it or not. So, as everything and everybody is changing around us, we can be sure that risk will be changing over time as well. We may consider a simple two-dimensional contingency model for this dynamic character of risk. Over time, risks are dependent on:

1   The ever changing *circumstances* – the *factual* and *objective* factors

2   The ever changing human *perceptions* – the *interpretative* and *subjective* factors.

While these factors change risk levels, their remediation measures may to have change as well. Over time, a certain risk may become more or less serious because of a mixture of changing objective and subjective factors, which are likely to influence each other as well. Risk management has to deal with these types of dynamics, preferably in the right direction of risk remediation and catching risk opportunities at the same time. These considerations stress the importance of a *cyclic* management of risk. Any changes in the identified and classified risks should be considered at regular time intervals. It brings us to *risk management*.

Risk management

Like most issues in business, risk also needs to be *managed* in order to stay successfully in business. Similar to the term risk, the term *risk management* also has many definitions. According to Clayton (2001), risk management is the overall application of policies, processes and practices dealing with risk. Risk management should therefore be a well-defined and understood responsibility within the entire project organization.

Many organizations suffer large cumulative losses from a myriad of apparently little incidents (Waring and Glendon, 1998). Sound risk management contributes to the direct and broader stakeholders of the organization, as well to society as a whole (Elliot et al., 2000). Thus, risk management is not only needed to avoid major disasters, but also to improve bottom line financial results. It is a pity, however, that these bottom line benefits of risk management are so difficult to measure. In the ground-related sectors of the construction industry in particular, there is hardly any available evidence about bottom line improvement by the application of risk management. In this respect, we should realize that risk management has been embedded, for years already, in many companies in other industries. These do consider their bottom line results as many of them are listed multinationals. Risk management would therefore at least pay back its investments.

Some research on the benefits of risk management in the construction industry has been conducted. According to research in the USA by the Construction Industry Institute (CII) of the University of Texas, a 1:10 cost–benefit ratio can be expected as a result of better contracting practices by improved risk allocation

(Smith, 1996). A study of two major transportation tunnel projects by Sperry (1981) resulted in estimated cost savings between 4 and 22 per cent, by the application of risk management practices.

In spite of these impressive figures, risk management remains primarily *management of expectations*. The word *expectation* derived from the word *spectare*, which means seeing. The word *ex* refers to something outside ourselves. So, expectation is something we see out there, even maybe something we anticipate in the future. It is possibly about managing the ground conditions on a construction site, by considering the effects of a number of ground profiles, some measured groundwater levels and a set of laboratory test results on pollution. Two generic approaches are selected from the variety in the literature to manage these expectations: heuristic and holistic risk management. These are particularly supportive to the introduction of the GeoQ ground risk management concept and therefore require some special review.

Heuristic or rule of thumb risk management

In this book I apply the so-called *heuristic* or *rule of thumb* risk management approach. It has a rather qualitative approach, which relies in particular on experiences and the collective judgement of individuals (Waring and Glendon, 1998). In addition, it fits well with the soft systems approach of engineering and construction.

Occasionally, this rule of thumb risk management approach will include some form of quantification of risk, which is the basis for the so-called *scientific approach* of risk management. This other broad school of risk management aligns with the hard systems approach. While it is normally based on complex mathematical and quantitative modelling, it is not as objective as it first appears. Any human individual performing this type of scientific risk analysis and risk management is inherently value-driven and influenced by his or her organizational culture. Current methods for quantified risk assessments are not capable of including typical soft system elements, such as power relations, motivations, organizational culture, as well as individual attitudes and perceptions. Therefore, at least initially, the heuristic or rule of thumb risk management approach is appropriate for most applications (Waring and Glendon, 1998). Detailed risk analysis, including statistical quantification according to the scientific approach, may be an appropriate next step with regard to the risk appraisal of technical (sub) systems. In ground-related risk management this approach proved to be viable and it serves consequently as the basis for the GeoQ concept of structured ground risk management.

Unfortunately, risk management requires more than just focusing on the major risks. As some kind of paradox, as already stated by Ansoff (1984) and Reason (1990), it is also very important that we pay attention to the so-called weak signals in the heuristic risk management approach. These weak signals may be important indicators of latent failures and risks (Smallman, 1996). They are often derived from sources of information that provide *qualitative* and rather *soft* or subjective information. It is all about balancing the right attention and resources towards the recognized important risks and these weak signals. The latter may accumulate to major risks later on, in the next phases of the project. It is this balancing of both the details and the bigger picture, within the unavoidable limits of resources such as budget, expertise and time, which makes ground risk management more of an art than just a rational mean. It keeps risk management, for any construction project, surprising, challenging, demanding and interesting as well.

Fatalistic and holistic risk management

Another set of major risk management paradigms can be distinguished: *reactive* or *fatalistic* risk management and *proactive* or *holistic* risk management (Smallman, 1996). *Reactive risk* management focuses primarily on risk *retention* and *transfer*. *Retention* of risk is, in fact, just accepting the risk with its consequences and losses, if the risk occurs. *Transfer* of risk requires another party willing to bear the risk. This reactive risk management approach is difficult and dangerous, because risk forecasting is limited by its inherent uncertainty, as well as by individual and team perceptions. This reactive risk management approach is more of a *laissez faire* of the recognized risk, where we or another party bear the consequences, if the risk occurs.

It is therefore often favourable to consider risks and their interrelationships on a more *proactive basis*, by considering potential risk and taking measures to *do* something to reduce the cause or effect of that risk. This approach serves as basis for *avoidance, prevention* and *reduction* of risk. Risk *avoidance* implies taking such measures that the risk is no longer present, such as the application of a piled foundation to avoid the settlement risk

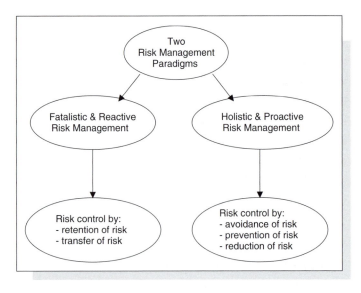

Figure 3.4: Fatalistic versus holistic risk management.

of a shallow foundation on soft soil. The *prevention* of risk is defined as taking measures to reduce the main causes and probabilities of occurrence of the risk. In the case of the shallow foundation, ground improvement will minimize the risk of settlement. *Reduction* of risk means taking measures to reduce the effects or consequences of the risk. In case of the shallow foundation on soft soil we may take certain measures to allow for minimal settlements without causing damage. Figure 3.4 summarizes the main characteristics of both described risk management paradigms.

Finally, Murphy's famous laws, like anything that goes wrong will go wrong, are also valid in construction. In any construction project many things can go wrong in numerous ways and most things can go right in just a few ways. A big pitfall of risk management is getting drowned in a pool of negativism, which brings us and all other participants in the risk management process into a negative vortex. It may easily become a negative spiral of all what can go wrong in our uncertain and dark world below ground level. Such a situation is not conducive to a positive attitude towards effective ground risk management. My approach is to emphasize the *bright side of risk*, to see risks, besides being threats as also opportunities to make a positive difference, in line with the presented holistic and proactive risk management approach.

The concept of ground and risk

Ground conditions

The theme of this book is managing risk as caused by differing and unforeseen ground conditions. These ground conditions should be considered in the widest context. Besides *ground* (soil, rock and all types of natural intermediate materials, such as collapsing soils, residual soils and weathered rock), *groundwater* is also considered here as an integral part of ground conditions. In addition, *pollution* (of water and soil) and *man-made structures* (such as pipelines, archaeological remains and old piles) are also part of ground conditions, as these prove occasionally to contribute significantly to ground related risk.

Ground risk management should manage all sorts of risk caused by these four main types of ground conditions. In other words, all uncertainties about the *presence* and *nature* of ground, groundwater, pollution and man-made structures belong to the scope of ground risk management.

Managed risk or wild gamble?

In 1992, the Associated General Contractors (AGC) of America provided a video, titled *Managed Risk or Wild Gamble – Getting on the Team*. It reflects that we have a *choice* about how to deal with ground conditions in our project. Do we gamble with the anticipated ground conditions? Following Brandl (2004), do we really want to play the game of *geo-poker* or *geo-gambling*? Of course, factors, such as severe competition on lowest price only, time pressure, a lack of knowledge, experience and awareness about the risks and dangers of ground conditions all provoke these gambling games. If we do so, then is it the obvious choice? Is it the only choice we have? Do our clients indeed expect us to gamble with the ground conditions, within their precious construction projects? In my opinion, we should manoeuvre a choice to *manage* ground conditions rather than *gamble*, for instance by using ground risk management. In order to do so and to develop a rationale for it, let us first consider some general characteristics about the unavoidable ground conditions for any construction project. Why does the ground surprise us so often in an unfriendly way?

Ground uncertainty and costs

Some statements, which you may encounter on a regular basis about ground conditions, are those like:

- The only certainty about ground conditions is its uncertainty

- Ground investigations – pay now or pay a lot more later.

These statements describe two main aspects of ground conditions that most of us will recognize: *uncertainty* and *costs*. Ground conditions can literally make or break our construction project. We may think of severe differential settlements, embankment failure, and so on.

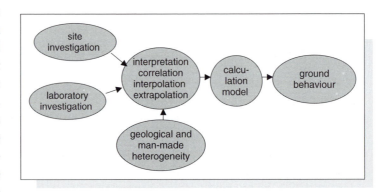

Figure 3.5: A simplified process model for dealing with ground conditions.

We have already explored the *random, fuzzy* and *incomplete* character of ground uncertainty. In addition, mistakes and incorrect interpretation of the often large and conflicting number of ground data are an additional source of uncertainty. Aspects such as the geological heterogeneity of ground and its non-linear and time dependent behaviour in relation to loads add further to the complexity of ground conditions. Furthermore, there is also heterogeneity in human perceptions of ground conditions, the interpretation risk, which is at least as difficult to tackle as the geological heterogeneity. How can we deal with all these uncertainties? Figure 3.5 presents a simplified process model for it.

Figure 3.5 can be used for all sorts of ground-related design and construction, such as a foundation design and an environmental clean up of a landfill. While hiding the complexity of ground conditions, this model provides a *forecast* of ground behaviour. The reliability of this forecast will surface during the construction and operation phases of a project. Then it will become clear to what extent the actual ground behaviour will agree with the forecast.

The mentioned uncertainties of *randomness, fuzziness, incompleteness,* as well as the added *incorrectness* are present in each step of any construction dealing with ground. *Information risks* are particularly relevant with regard to ground investigations, while *interpretation risks* play a dominant role in the design process. Within ground-related engineering, the accumulation of these uncertainties results in an overall uncertainty of at least 50 per cent. This is reflected in the widely accepted and applied overall safety factors of two, three or even more for geotechnical design. This overall uncertainty for ground is much higher than for other construction materials, like steel and concrete, with respective material uncertainties of some 5 per cent and 10 per cent (Wentink, 2000). By now, we should be convinced that ground uncertainty is a given fact and costs have to be made to reduce it to an acceptable level within the boundary conditions and risk profile of the project. However, the hybrid character of ground risk makes it even worse.

The hybrid character of ground risks

A complication with regard to ground risk is its hybrid character. Let me explain this statement with an example. Main risk types for construction projects are *external risks, organizational risks, technical risks* and *legal risks*. Occurrence of any of these risk types will have financial consequences and may affect other project success factors such as safety, quality, planning and reputation. Table 3.1 presents these four main risks types with their potential ground-related risk categories, in a summarized and slightly adapted version of a risk identification list by Altabba et al. (2004).

Table 3.1 shows the external, organizational, technical or legal aspects of ground-related risk. Given the high potential impact of ground-related risks, it is highly advisable to consider ground risk as a *separate* risk type, in order to avoid confusion and undesired mixing with other risk types. This will ensure that the inherent hybrid ground risks get the adequate attention they need.

Main ground risk types

The previous sections introduced the concepts of uncertainty, risk, risk management and ground conditions. Before entering the concept of GeoQ ground risk management, I will focus in some detail about *what* are in fact ground risks.

Table 3.1: Some risk types and their potential ground-related risk categories

General risk types	Ground-related risk categories
External	
Natural hazards	Ground characteristics
Environmental	Soil/water condition and or contamination
Organizational	
Schedule delays	Unforeseen site conditions
Cost overruns	Contractor claims
Technical	
Design	Inadequate ground data and or ground characterization
Construction performance	Differing site conditions
Legal	
Contractual	Misinterpretations
Third party	Environmental aspects

We have already recognized their hybrid character, but what type of risks can and should we actually classify as ground risks? Box 3.8 presents the four main elements of ground in relation to their equivalent risk types.

Box 3.8   The four main elements of ground and their related risk types

1   *Ground* (soil, rock and all types of natural intermediate materials, such as but not limited to collapsible soils, residual soils and weathered rock): *geotechnical risk*

2   *Groundwater* (of natural origin, which can be sweet, brackish or salt): *geohydrological risk*

3   *Pollution* (of water and soil): *environmental risk*

4   *Man-made structures* (such as pipelines, archaeological remains and old piles): *man-made obstruction risk.*

The four main risk types of Box 3.8 are more specifically described by a number of examples in Box 3.9.

Box 3.9 A brief introduction to a number of ground-related risk types

1 *Geotechnical risks*

- Deformation – differential settlements

- Strength – instability, collapse, hard ground layers.

2 *Geohydrological risks*

- Unexpected groundwater tables

- Groundwater overpressure, perched groundwater tables

- Fast changing groundwater conditions due to weather influences.

3 *Environmental risks*

- Polluted groundwater

- Polluted soil.

4 *Man-made obstruction risks*

- Archaeological remains

- Old piles

- Electricity cables, water, gas, oil pipelines in operation.

Obviously, the listing of Box 3.9 is not at all intended to be complete. The presentation of detailed and project specific risk registers is clearly beyond the scope of this book, with a focus on the *processes* of risk management. Risk registers of various types and complexity and can be found in the literature and on the Internet. Clayton (2001) presents a number of risk registers, which were retrieved from some large construction firms in the UK.

Finally, Box 3.10 shows a few ground-related risks in some more detail. According to the previous definitions, these risks are divided into two parts, the *probability* of occurrence and its *effects*, in case of occurrence.

The risks, as presented in Box 3.10, are all foreseeable and mainly pure risks, as expressed in our previously applied terminology. If these risks actually occur, they have in common a negative effect on project success factors like budget, planning, quality and even reputation of the parties involved. The risk of the last example of Box 3.10 has a somewhat different character, it is a typical foreseeable and speculative risk, with an opportunity of cost reduction. After these classifications and examples of typical ground-related risks we are now ultimately ready to enter the next section: the GeoQ concept.

Box 3.10   Some examples of ground related risk by probability and effect

1   *Pile fracture* by an unexpected hard sand layer or presence of boulders – the ground-related risk includes the *probability* of pile fracture and the *effect* of a broken pile in terms of additional costs and delay

2   *Differential settlements* by a locally present soft peat layer – the ground-related risk includes the *probability* of the presence of a soft peat layer and the *effects* of too large differential settlements in terms of damage, like serious cracks in the structure and the associated costs of repair

3   *Instability of an embankment* by liquefaction of an unexpected loose sand layer, with an apparent normal cone resistance – the ground-related risk includes the *probability* of that unexpected loose sand layer and the *effect* of embankment instability with serious safety concerns, due to injured or even lost workers

4   *Collapse of the bottom of a construction pit* due to an unexpected groundwater pressure in the underlying sand layers – the ground-related risk includes the *probability* of deviating groundwater levels and the *effects* of a drowned building pit in terms of damaged excavation equipment, delay and its associated costs

5   *Unnecessary expensive dike design*, within the project boundary conditions, because of over-conservative design parameters. The ground-related risk includes the *probability* of stronger ground than assessed and the *effect* of an over-expensive dike design, which implies the client pays too much and the contractor makes unnecessary costs and loses some (additional) profit.

The concept of ground risk management

What about the concept of ground risk management? Still, the practical and structured dissipation of ground risk management is rather slow. The concept is by far not fully recognized and a lot of its benefits remain untouched. The main reason for the current situation is probably the *gap* between the rather abstract and general theory and its application. A practical framework for ground risk management has been awaited. Furthermore, ground-related risks are often seriously considered in only one or two specific stages of a project. In the pre-design phase, risks may be analysed thoroughly, even with sophisticated tools like Monte Carlo analyses. The results are reported, often together with a spaghetti of cause and effect figures. However, this type of risk analysis is not easily accessible and applicable for those involved in the actual and later stages of the project. Consequently, the often-expensive risk report will be stored in the darkest corner of the office-desk and during the design and construction phases no follow up takes place. In other words, the ground risks are not *managed*.

The GeoQ concept

The GeoQ concept forms the beating heart of this book. It was initiated in the year 2000 at GeoDelft, the Dutch National Institute for GeoEngineering in The Netherlands (see Box 3.11).

Box 3.11   GeoQ or how it started

During a meeting in the year 2000, Jacob, at that time a project director of Schiphol International Airport, The Netherlands, challenged me. We discussed the common practice of presenting geotechnical advice. 'I do not want *one* single settlement prediction', he stated. As an experienced and professional client, he knew that settlements in reality always differ from their calculated forecast. Jacob continued: 'I want to see a certain *range* of possible settlements, together with their probability of occurrence and preferably expressed in sound percentages'. Only then, as a professional project director, he would be able to decide, together with his project team, about any of the most appropriate settlement reducing measures. He would be able to balance risk and cost consequences, according to his own preferred risk profile, as he is ultimately responsible for the project budget. During that meeting we decided to provide a few calculations in his proposed setting. We also decided to evaluate one of his projects, an infrastructure project. The objective of the evaluation was to check whether more ground data, earlier in the project, would result in better project risk management. It resulted in a structured and cyclic risk driven approach, which evolved towards the GeoQ concept.

GeoQ is a risk-driven attempt to challenge and innovate the conventional ground engineering and construction processes. This new process emerged in an evolutionary way, together with the ground engineering and construction industry, by application of existing risk management principles in a number of stages in construction projects. With GeoDelft in a coordinating role, many innovation-driven parties in the Dutch engineering and construction industry, including clients, engineers and contractors, applied elements of GeoQ in their projects. This resulted in a variety of GeoQ cases, as described in the remaining part of this book, which became available over 5 years. This mixture of existing and generic risk management theory and its practical translation and application in the geotechnical parts of construction projects resulted in the actual GeoQ concept.

GeoQ, with the *Q* of *Quality*, is a *risk-driven* approach to manage ground conditions and behaviour in a structured way, in order to arrive at a successful project for stakeholders. It is a *cyclic* risk management process as well. It makes the requirements for ground data at every stage in the project, in terms of the type, quantity and quality, more explicit, in order to create a true *quality* construction project. This desired type of project moves beyond meeting safety and quality standards. It is realized within *budget* and *planning* as well, which strengthens the *reputation* of the parties involved. The GeoQ process provides a straightforward ground risk management framework, which is flexible enough to be adaptive to the specific requirements of each small or large construction project. It is applicable for any ground conditions, anywhere in the world. Box 3.12 presents the three main characteristics of GeoQ.

Box 3.12   The three main characteristics of the GeoQ process

1 *Cyclic* risk management, by the repetition of six risk management steps in six generic project phases and the structured storage of its results in accessible and easy-to-use risk registers

2 *Just in time* availability of adequate ground data, driven by the perceived and agreed risk profile of the project. In the early project phases general ground information might do, in further project phases more detailed ground information should be used

3 *Continuity* of adequate ground data, which indicates the need for permanent and well-structured access to all available ground data, including that from the preceding project phases. Ground data should be stored in databases and any identified risks in risk registers. This information needs continuous updating with new insights during the subsequent project phases.

Possibly surprisingly, there are many GeoQ *tools* already available in the global ground engineering and construction communities. These tools can greatly contribute to an effective and efficient GeoQ process. They vary from ground risk identification and classification methods, via advanced ground investigation equipment to geotechnical software for probabilistic geotechnical design. These tools have already been widely and successfully used for many years in our practices. However, they will have an interesting *additional* value, when applied in the risk driven context of GeoQ, because of their contribution to explicit ground risk management.

The intention behind GeoQ is to serve all project stakeholders, i.e. the client, the engineer, the contractor, the public, as well as suppliers or partners and the employees of the involved organizations. Ideally, GeoQ is applied throughout the entire project lifetime. It starts at the feasibility phase and continues during pre-design and design, the contracting and construction phase to the operation and maintenance phase.

In 2001, a number of countries of the European Union introduction the euro with their ultimate statement: 'The Euro is for all of us'. GeoQ is for all of us as well, even well beyond the boundaries of Europe. Everybody, anywhere in the world, is free to use the GeoQ principles in his or her projects. GeoQ can be perceived as an *open source ground risk management framework*. It is comparable with other worldwide open-source developments, such as the Linux software, the Apache web server programme and the Wikipedia Internet encyclopaedia. GeoQ can be adapted, as requested by specific requirements of projects and stakeholders, so long as the *core objective* of GeoQ is concerned: serving the overall quality of a construction project by structured ground risk management. It should serve as our ground for success!

What are the main benefits of this structured ground-risk management concept? GeoQ helps to manage *foreseeable* ground risks in an effective and efficient way. It provides *proactive risk management* and has therefore to start in the earliest stage of a project to maximize its benefits. In this stage, many options for design and construction are still open and to be decided upon. It means that an adequate ground risk response can still be included rather smoothly, without major design or construction modifications, with all their often unfavourable costs and construction time consequences.

The GeoQ process supports a thorough insight into a lot of interrelated ground aspects and their associated risk. It facilitates management of the risk of ground, groundwater, pollution and man-made aspects, such as buried cables, in an integrated way, which minimizes any suboptimum solutions. Furthermore, it relates ground aspects to other technical and non-technical project risks and vice versa. It helps to demonstrate the need for timely and adequate ground attention in order to manage the major overall project risks. The application of GeoQ justifies ground investigations to non-geotechnical decision-makers of a project, because its added value can be communicated within a risk management objective by relating the expected ground investigation benefits to meeting the overall project objectives. GeoQ proves to be of great help to convince slightly sceptical clients about the importance of adequate ground investigations with their associated investments.

Furthermore, in any construction project, the contractual phase needs special attention, especially when it comes to the widely underestimated importance of clear contractual statements about differing ground conditions. The *allocation* of all identified risks to the parties involved is of major importance during contracting. Each identified risk needs one or more owners in order to guarantee effective management of that risk. One of the benefits of the GeoQ approach is its transparent management of the contractual consequences of differing ground conditions. Within the GeoQ process, *risk allocation* is provided by the so-called *Geotechnical Baseline Report* (GBR) method, which was originally developed in the 1990s in the USA (Essex, 1997). This geotechnical baseline approach is entirely embedded in the GeoQ approach. The GeoQ process is well suited for both traditional and modern innovative types of contracting, such as design and construct. So far as I know, a similar approach to this is not yet available.

In summary, the GeoQ process allows a certain degree of *risk space* in our construction projects. It provides sufficient *flexibility* to allow for changing risk perceptions and project circumstances. Normally, the risk space shrinks during the project as a result of the information and experience gained. However, ground has its inherent uncertainty and may expand the risk space again at an unexpected moment. GeoQ ground risk management helps to facilitate this continuous breathing of any construction project in its risk space and serves the required flexibility for adoptions and change, at any time.

Finally, it needs to be said, the ground situation as managed by the GeoQ process is a *part* of the entire project management. GeoQ deals with ground and, if the remaining non-ground-related aspects are not given equally sound and structured management attention, GeoQ will not save the project. We will, therefore, next explore the positioning of GeoQ in a number of existing ground-related disciplines and project management practices.

GeoQ and the ground-related disciplines

What is the position of the GeoQ concept in relation to the conventional ground-related engineering and construction disciplines? What about the existing risk management with regard to GeoQ? What about natural hazards and GeoQ? What is the position of GeoQ towards the well-established quality management and knowledge management areas? In this and the following sections, the GeoQ process is positioned relative to a number of existing disciplines and management approaches. There is one overall message: GeoQ intends to *complement* existing disciplines, knowledge and experience, rather than *compete* with or replace them.

As an engineering geologist, I like the fact that both Karl Terzaghi, the founding father of soil mechanics and Leopold Müller, the founder of rock mechanics, considered themselves as engineering geologists. Together with, for instance, Bock et al. (2005) I regret, however, they themselves and their successors are not yet successful in establishing engineering geology as a so-called independent discipline, with its recognition in our engineering and construction community.

This is just an example of a tendency to think from our own background and disciplines and put that viewpoint in the centre of the world. Perhaps for that reason I recognize a lot of confusion, both in the literature and in practice, in several terms used and definitions of applied geosciences. Just to mention some: *soil mechanics, rock mechanics, soil engineering, rock engineering, geological engineering, engineering geology, ground engineering, geotechnical engineering, environmental engineering* and *geoenvironmental engineering*. All of these disciplines are related to each other, with ground as the main binding element. The introduction of a new discipline, ground risk management, might even increase this confusion. To avoid this, I like to use the structure of the applied geosciences, such as for instance indicated by Bock et al. (2005). It is not an astonishing brand new structure, but it facilitates a clear positioning of ground risk management. In this structure *soil mechanics, rock mechanics* and *engineering geology* are considered as the three pillars of *ground engineering*. I hope that most readers can accept my simplification when considering ground engineering and geotechnical engineering as equivalent to each other. To these three disciplines I add the recently developed discipline of *geoenvironmental engineering*. The latter I define as the discipline dealing with polluted ground and groundwater during all engineering, construction, and operation and maintenance phases of any construction project or area. In-situ cleaning of a polluted site is, by this definition, a typical example of a geoenvironmental project. Too often, I recognize that geoenvironmental engineering is considered as a separate discipline, with rather isolated codes and practices, separate ground investigations and consequently a lot of suboptimizations. Is it really effective to have first a geotechnical site investigation, followed by a geoenvironmental site investigation just a few weeks later and often provided by a different contractor and a different engineer under a different contract? Environmental engineering still seems not to be fully integrated into the entire construction project. Such an integration will give lots of opportunities for creating synergies because of the high interdependence of construction and adequate dealing with environmental issues.

In summary, for the purpose of this book I define ground engineering and construction as the discipline that deals with any type of engineering and construction of any structures in, at or with any type of ground. In this definition *soil engineering, rock engineering, engineering geology,* and *geoenvironmental engineering* are all considered as being part of ground engineering, as presented in Figure 3.6.

It is probably no coincidence that the three relevant International Societies representing soil mechanics, rock mechanics and engineering geology, are moving step by step towards each other, in order to integrate finally into some kind of federation of geo-engineering societies (Knill, 2003). Furthermore, I consider ground engineering as a *facilitating* and *supporting* discipline for the entire construction industry, as any structure will be built on or in ground. Castles in the air are beyond the scope of this book.

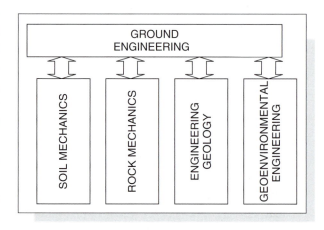

Figure 3.6: A simple framework of the four main disciplines of ground engineering.

Ground risk management is defined here as management of the uncertainties about the presence and nature of ground within ground engineering. As I mentioned before, by the introduction of GeoQ ground-risk management I do *not* intend to introduce a new isolated discipline. On the contrary, I would like to develop it towards an integral and inevitable part of risk- and opportunity-driven ground engineering. The GeoQ ground risk management concept is by no means meant to *replace* the conventional disciplines of soil engineering, rock engineering, engineering geology and geoenvironmental engineering. It is a fear of ground-related professionals that I sometimes encounter when introducing GeoQ. These disciplines have proven already for many years their value to our societies. We need all this deep knowledge and experience alive and kicking. GeoQ ground risk management should *add* and contribute to these proven disciplines and to ground engineering as a whole. GeoQ intends to provide a strong connection between the technological approach of the mentioned disciplines and the dynamic and complex outer world of our societies, where the support for purely technological disciplines and solutions is diminishing. GeoQ can bring these disciplines, together with their experts, back into the spotlights of society, as the risk driven approach will highlight their value to society. In fact, GeoQ can be considered as some sort of ground-related *communication vehicle*. It may help engineers to become better understood and appreciated by the many stakeholder groups in society, such as policy makers and the public. Many of them may have a rather limited technological awareness, but they have unmistakable high interests in the final results our engineering and construction projects deliver to them.

GeoQ and risk management

The GeoQ method is quite new. However, considered from a distance, it will quickly become clear that many of the elements of GeoQ are concepts that have been well known and proven for some time. In fact, GeoQ is literally a deepening of existing risk management procedures, such as *MARIUN* (MAnaging RIsk and UNcertainty) in the UK and *RISMAN* (RISk MANagement) in The Netherlands. The latter has been in use since 1995 and successfully applied to many large and smaller infrastructure projects.

However, a detailed and fit-for-purpose translation of these systems towards management of ground risk, with its own characteristics, was still missing. In practice, systems like RISMAN proved to be more of a one-moment risk *analysis* tool, rather than a risk *management* tool for the entire construction process. During the early application of the GeoQ principles in construction projects, many colleagues and I occasionally experienced some hesitation. For instance, in the mind of an engineer, something might be easily considered as wrong if it does not prove to be completely right. It is a way of *digital thinking*: it is one or zero. This type of black and white approach is often very useful for the technical solutions within construction projects. However, the realization of the *construction project* includes a *construction process*, from the early feasibility stage via pre-design, design, construction towards operation and maintenance. It includes both the hard technical and the soft human systems. Blockley and Godfrey (2000) taught us that optimizing a construction process by using solely black and white thinking is not productive at all. Simply because it disturbs the inherent subtle and dynamic character of the construction *process*, with all its internal and external influences. Therefore, I have to disappoint those readers who might expect a *one size fits all approach* for their projects, by the application of GeoQ. Such an over-simplistic risk management approach cannot exist given the uniqueness of each construction project. I compare it with the two most valuable words that I learned while taking my MBA: *it depends*. When positioning GeoQ in today's risk management landscape, we permanently need to remember these two words.

At this stage it should be noted and taken seriously that the application of ground risk management systems such as GeoQ, similar to any system, is not a definite guarantee that nothing will go wrong within our construction projects. GeoQ is just a few years old and it has the characteristics of a young child. It is discovering the world and its place in that world. A lot of GeoQ practices and tools will probably be further developed in the next years and even decades, based on newly gained GeoQ experiences. To conclude, GeoQ is literally a deepening of existing risk management procedures. Its further development may be highly facilitated by further importing and translating theories, practices and lessons learned from risk management in any other sector or discipline.

GeoQ and natural hazard management

Large parts of our world are increasingly vulnerable to the effects of *natural hazards*, such as earthquakes, tsunamis, floods, hurricanes, eruptions of volcanoes and landslides. These have a major negative impact on the people living in the affected areas, with serious losses of lives, infrastructure, goods and services. In addition, many areas in the world are regularly affected by less dramatic but still significant ground-related problems, such as those weather-related ones seriously affecting the railway industry in the cold regions. As an example, much of the tens of thousands kilometres of railway in Canada are susceptible to embankment failure, as a result of rapid melt of snow or drainage disruption. The annual financial losses because of derailments and delays amount to millions of dollars (Gitirana and Fredlund, 2005). Another indication of the impact of natural hazards on society is presented in Box 3.13.

Box 3.13   The impact of natural hazards on society

To illustrate the impact of natural hazards on society, I looked at some figures about Latin America and the Caribbean provided by the Inter-American Development Bank. In the period between 1975 and 2002 the average *annual* losses caused by natural hazards were estimated at 3 200 000 million US dollar. In that period 250 000 deaths were recorded, according to Mora and Keipi (2005). I am afraid that similar figures can be found for other parts of the world with a relatively high natural hazard risk.

   The same authors, Mora and Keipi, are of the opinion that, in Latin America and the Caribbean, existing decision-making processes are unfavourable for the prevention of the effects of natural disasters. Instead, centralized and reactive disaster management systems typically prevail. It should be recognized here that the opinions of Mora and Keipi do not necessarily reflect those officially adopted by the mentioned Inter-American Development Bank. As a frequent reader of newspapers, I am convinced that this situation also may occur in many other parts of the world.

In my opinion, the situation as presented in Box 3.13 calls for the inclusion of sound and relevant risk management criteria throughout the entire cycle of construction projects in order to include the potential effects of natural hazards on the particular project in the design process. This is expected to result in a higher degree of natural hazard awareness, as well as easier ways of understanding the effects of natural hazards on a particular construction project. Obviously, it will not prevent any single natural hazard from happening. It may, however, reduce the negative *effects* of the natural hazard on the constructed objects and its stakeholders. I understand that there may be serious budget constraints in doing so, but the application of risk-driven concepts, such as structured ground risk management, does not automatically require more budget. It is a new way of perceiving and operating, within the inevitable budget constraints of any project in any part of the world.

The GeoQ ground risk management concept does absolutely not intend to *replace* the various existing risk management systems and procedures for assessing and correcting the effects of natural *hazards*. Hazard management covers occasionally (large) areas rather than just project sites, which can be covered by techniques such as *hazard zone mapping*. Based on these hazard maps, an even more suitable construction site might be selected or appropriate site preparation measures can be undertaken. Given these known hazards, the GeoQ process can be used to minimize the potential harmful effects of these natural hazards on the project, from a ground conditions perspective. GeoQ is clearly *complementary* to existing natural hazard management concepts and methods, by appraising and above all, *anticipating* the potential harmful effects of natural disasters within the design, construction and maintenance phases of construction projects.

GeoQ and quality management

GeoQ bears the letter Q of quality, which is obviously not a matter of coincidence. Quality management is an established and an inevitable part of project management in the construction industry. In this section GeoQ ground risk management is discussed in the light of quality management.

Many experts, including the three famous American gurus, Phillip B. Crosby, W. Edwards Deming and Joseph M. Juran, define quality in various ways. In basic terms, quality is simply meeting the requirements of the customer (Oakland, 1993). It is therefore not possible to provide quality without knowing the requirements of the client. The same applies for risk management – effective risk management is not possible without including the risk requirements and risk tolerance of the client, while risk management goes even a step further. In a real quality project, the risks of all stakeholders in the project, not only the client, are managed in an effective way.

The need for risk management appears to agree quite well with the general considerations and objectives when applying quality management. The key reasons for quality management are more (global) competition, more demanding customers and the never ending need for market share and cost control (Dibb et al., 1997). Therefore, in today's highly competitive and fast changing business environment, quality is not only required for profitability, but crucial to business survival (Muhlemann et al., 1992). Furthermore, like most issues in business, such as risk, quality also needs to be managed. Quality should be a well-defined and understood responsibility within the entire organization, including senior management, marketing, research and development, production/operations management, etc. (Muhlemann et al., 1992). One aspect addressing this is a fully documented quality system, for instance according to ISO 9001 standards, which will support meeting the customer (and thus quality) requirements, while also meeting the objectives of the organization. That is, at least according to the theory, as illustrated by the example in Box 3.14.

Box 3.14   Quality management and ground investigations

According to my international experience, particularly regarding ground investigations, the *practice* of meeting the customer requirements is often quite different from the proposed quality management *theory*. Many customers simply assume quality as a (paper) given and solely select their ground investigation contractor on the lowest price criterion. Due to severe price competition, it seems to be almost impossible to stay competitive *and* to keep quality to minimum standards for these contractors.

For instance, the introduction of an industry-wide quality system for cone penetration testing (CPT) in The Netherlands has failed, to date. Although a Dutch CPT standard with four distinctive quality levels is available, in line with international recommendations, the differences in CPT quality are still fuzzy in practice. CPT providers have not yet succeeded in communicating the benefits of CPTs with distinct quality and, consequently, clients are not willing to pay a price premium for higher quality CPTs. Another example concerns soil drilling and sampling. Serious soft soil sample disturbance, which I occasionally encounter in soil laboratories, is also a silent witness of still rather underdeveloped quality management within the ground investigation sector.

As Box 3.14 may teach us, quality management should ideally be supported by an awareness of the costs of (lack of) quality. The same applies for risk management!

Total Quality Management (TQM) is a step beyond sole quality management and is widely recognized as a concept that emphasizes continuous improvement, product quality, customer focus and empowered employees. Results of TQM programmes can be staggering, such as a 40 million US dollar reduction of the Ford Motor Company operating budget by adopting total quality principles and changing corporate culture (Daft, 1998). As in the case of quality, a general definition for TQM does not exist (Lowery et al., 2000). TQM is even considered as some form of pre-1965 Japanese Total Quality Control (Dean, 1998) and here we rediscover the cyclic character of TQM, which is similar to the cyclic character of the GeoQ process.

Indeed, there are striking similarities between quality management and risk management. In fact, it appears that GeoQ has a lot in common with the Japanese quality management concept of Kaizen. This quality concept is also a cyclic and a process-oriented step-by-step approach in which everybody is involved (Imai, 1986). This rather collective approach towards quality management differs from the conventional western style, which is basically product- and result-oriented. It may easily neglect the process and quite often excludes real participation of all related people, because of some kind of top-down command structure. The reward of effort made, and not only results, is another characteristic of Kaizen. In addition, it has a positive and forward-looking approach. Kaizen assumes, as a given, that every organization has its problems, at least the so-called *warusa-kagen*, things that are not yet problems but are still not quite right (Imai 1986). Within Kaizen quality management, every problem is perceived as an opportunity for improvement. Similarly, every risk identified by the GeoQ ground risk management process is still a risk and can be managed in some way or another. The GeoQ concept appears generally to agree with the quality circle philosophy, which Deming introduced in Japan in 1950.

So after all these similarities between risk management and quality management, are there any differences between these two types of management? Why do we need them both? We need them both, because they *complement* each other. It is believed that the founding father of quality management, Deming, said that the more important something is, the more difficult it is to measure. Measurement, which is in fact the ultimate way for quality control, excludes, however, what we cannot yet measure. It excludes what we do not ultimately know yet. Quality management excludes hypothetically thinking about what is still uncertain. A fully documented quality system, for instance according to ISO 9001 standards, intends to meet the customer (and thus quality) requirements, while also meeting the objectives of the organization. Quality management may even be supported by the awareness of the costs of (lack of) quality and the application of tools such as statistical process control. However, dealing with risk and uncertainty is *not* part of the quality system, while this will have a major impact on meeting the quality standards. That is where risk management can help us. Risk management is not solely about measuring, in spite of the fact that there are a lot of quantitative and semi-quantitative tools. Risk management can be used to map the unavoidable uncertainties of the project. It goes beyond known and foreseen certainty and enters the area of foreseen uncertainty and even unforeseen uncertainty. Risk management starts where conventional quality management stops.

In a summarizing conclusion, quality management is about realizing the project within the pre-set quality standards, while risk management is dealing with risks of not applying these pre-set quality standards. Regarding ground conditions, GeoQ ground risk management covers the area *outside* the pre-set quality standards of quality management. This is the area where GeoQ is positioned when it comes to quality management.

GeoQ and knowledge management

Finally, while arriving at the end of this chapter, I feel the need to dedicate a few sentences to the position of another type of management: *knowledge management*. What is the position of ground risk management in respect to knowledge management? In a simple definition, knowledge management concerns management of what we know in our (project) organizations. Whether knowledge is theoretical or practical, there are limits to our knowledge, it is always finite. Risk management in general and GeoQ ground risk management, in particular, face uncertainty, they try to reveal what we do not know (exactly) yet. Risk management considers what may happen when we use our imagination, which is infinite. Therefore, risk management primarily asks questions, like *what if*, while knowledge management basically provides answers, like *if then*. Knowledge management and risk management seem to complement each other on the continuum from knowing towards not knowing. In other words, it brings us from fact to fiction, and the latter may become a fact in the future. The more facts or knowledge we have, the

---

Box 3.15   Knowledge meets risk management by GeoBrain Foundations

The development of Internet accessible experience databases started in The Netherlands around 2002 by the so-called GeoBrain Foundations. It is an on-line experience database, in which the top ten foundation contractors in The Netherlands store their non-competitive foundation experiences. The database is operated by GeoDelft, the Dutch National Institute for GeoEngineering. The purpose of GeoBrain is to provide society with all relevant knowledge and experience in a structured way. It aims to provide engineers with local and practical ground experiences, which they may use to verify the feasibility of their theoretical designs. For instance, if you have to design a sheet pile wall in the city centre of Amsterdam, then you can access the database and view the existing experiences in that area, just by searching under the area's postal code. It is an easy and quick method for a design check with local experience data and can be considered as a viable risk management tool in the design phase of a construction project (Bles et al., 2005).

more effectively we can deal with this fiction in terms of its relevance for decision-making. That is why knowledge management may become an interesting partner of risk management. One example is the increasing use of risk registers and large experience databases, which may become available via Internet, as presented in Box 3.15.

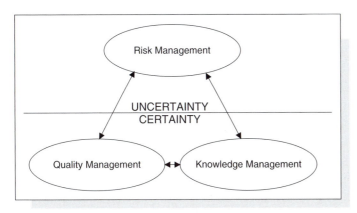

Figure 3.7: Partnering of risk, quality and knowledge management.

In conclusion, risk management provides methods and tools to deal with the 'not knowing part' of the knowing–not knowing continuum. Risk management in general and management of ground-related risk, in particular, intend to complement, to add towards knowledge management, instead of replacing it. Just as in the case of quality management and illustrated in Figure 3.7, risk management should be considered as a partner of knowledge management, rather than a competitor for the always too limited resources for operating these types of management.

It is up to the dedicated managers in charge to distribute their resources in the most effective way towards the complementary disciplines of quality management, knowledge management and risk management.

Summary

The concepts of uncertainty, risk, ground and GeoQ appear to serve as a sound foundation for the structured management of ground-related risk. Three types of uncertainty help to recognize the inherent uncertainty of ground: randomness, fuzziness and incompleteness. A number of examples have demonstrated that all three of these are abundantly present in our day-to-day ground engineering and construction practice.

From the many available definitions, a very simple and practical risk management definition fits our ground risk management purpose: risk is the product of the probability of an uncertain event and the consequences of that event. Uncertain events can be either negative, creating problems, or positive, creating opportunities. With regard to ground engineering and construction it is helpful to distinguish between certain risk types, such as pure and speculative risks, foreseen and unforeseen risks and information and interpretation risks. A complicating factor is the dynamic character of risks: they change over time.

The heuristic or rule of thumb risk management approach is well applicable to manage ground-related risks, because it allows the combination of factual data with experiences and judgement of individual professionals. It fits well with the random, fuzzy and incomplete character of ground. Furthermore, risk management needs a proactive and holistic approach, which is primarily focused on the avoidance, retention and reduction of risk on a proactive rather than reactive basis. The holistic approach considers risk not as an isolated and pure technical issue, but closely related to its social environment. The inherent subjectivity of risk and risk management has been highlighted as of primary concern.

The danger of gambling with ground conditions, which often results in an adverse relationship between ground uncertainty and costs, has been illustrated by the concept of ground conditions and risk. Ground risks have a hybrid character, which supports the vision to consider them as a separate risk type in order to guarantee sufficient attention to this major risk source in construction projects. Four main types of ground-related risk have been identified: geotechnical risk, geohydrological risk, environmental risk and man-made obstruction risk.

The GeoQ concept for structured ground risk management aims to improve the overall quality of construction projects, in terms of optimizing costs and profits, planning, safety, functionality, sustainability and reputation. Three main characteristics of GeoQ are cyclic risk management, just-in-time availability of the required ground data and continuity in availability of that data, during the entire construction process, preferably by the use of database and risk registers. GeoQ is literally a deepening of existing risk management principles towards ground conditions. It complements the well-established and proven functional disciplines of soil mechanics, rock mechanics, engineering geology and environmental engineering. Furthermore, GeoQ is intended to complement, rather than replace, existing management disciplines, such as hazard management, quality management and knowledge management. GeoQ structured ground risk management is applicable in any construction project in any ground conditions. It is an open and adaptive framework, free of use by anybody, everywhere in the world.

PART TWO

The people factor in ground risk management

4 Individuals and risk

Introduction

> Choose a job you love and you will never have to work a day in your life
> Confucius, ancient Chinese philosopher

Ground investigations are an indispensable tool for managing ground-related risk, typically involving the exploration of hidden ground characteristics. They provide the information needed for any successful construction project. The application of risk management in daily practice requires change, both by ourselves and the people we work with. As an analogy to ground investigations, this chapter therefore aims to explore ourselves. Without usurping psychologists, experience has shown that it can be helpful to explore ourselves to some degree, before we start on the slippery path of change, uncertainty and risk. What can *we* actually do, when implementing ground-related risk management that may help to transform the industry?

We start by introducing the concept of the *individual*, followed by the concept of *perception*. Our perceptions often appear to be less objective and rational than we would like them to be. It is particularly important to realize this when considering individual perceptions and their effects on risk.

Finally, this chapter perceives our own contribution. What capabilities do we need to create irreversible and positive change? Definitive answers are not given, simply because they are not available. Was it Albert Einstein who stated that the answer is already enclosed by the question? However, six key-capabilities for the application of ground-related risk management are provided. This chapter again ends with a summary.

The concept of the individual

It is easy to talk about individuals, but what exactly does it mean? Can we define the term individual and recognize the importance of a word that is often unwittingly used? *Individual* literally means 'not to separate'. One valuable charac-teristic of an individual is authenticity. The core of authenticity is sincerity about ourselves. It implies an acknowledgement that we are who we are, and nobody else. Authenticity gives us the courage to be different (Kets de Vries, 2000, 2002). Such courageous individuals are vital for implementing new concepts, such as ground-related risk management.

Our discontinuities

Being an authentic individual proves to be not that easy. Like rock, we may be influenced by discontinuities. Those readers involved with rock engineering are familiar with the presence of discontinuities in rock masses. Discontinuities are a form of fracture that occurs in a regular pattern. Depending on their characteristics, such as orientation, spacing, and the opening, they can dramatically reduce the bearing capacity of rock masses.

We can also experience discontinuities in ourselves, when we are not sincere about ourselves. We often have to bridge the gap between our *personal self* and our *professional self*. Our work, for instance, may demand a fast and isolated geotechnical design check, while our inner feeling tells us that more time and budget are needed to reach a responsible opinion. In such a situation, do we work non-paid overtime? Have we already worked overtime almost continuously for the last three months as well? We struggle with feelings of authentic guilt and neurotic guilt (Block, 2002). The former is related to feeling insincere about ourselves. We are not authentic in that situation. The latter implies that we have feelings of guilt, because we are not fulfilling the expectations of others. For instance, not performing that 'quick and dirty' design check for a client.

Is this soft talk about hard rock? What is the problem with individual discon-tinuities? It is not simply because they may lead to a feeling of fracture, they may create a sense of unrest too. We may feel tired, which makes us less effective in our work. Discontinuities may affect our sincerity and stability as a person as well. In terms of rock mechanics, if the aperture of the rock discontinuity grows too wide, the rock mass becomes unstable or even collapses. Stress, sickness and burnout are the ultimate results of struggling too long with our inner discontinuit-ies, symptoms experienced by many professionals throughout today's world. The associated costs for their companies and society are high.

These gaps between our *personal* and *professional* selves are simply not effective from a business perspective. Our inner conflicts are often recognized by colleagues, employees and clients, who realize that there is a degree of insincerity within us. They may, in fact, feel that what we say and do is actually different from what we think and want. Trust is a prerequisite for change, as many textbooks about change management teach us. This lack of authenticity undermines the precious *trust* of the people around us in our intentions and actions. In the words of Goffee and Jones

(2005): 'If a leader is playing a role that isn't a true expression of his authentic self, followers will sooner or later feel like they've been tricked'. A substantial lack of trust will make it virtually impossible to implement any sort of change.

We need truly authentic individuals, preferably with a minimum of discontinuities. Only they will be able to create *authentozoic organization* (Kets de Vries, 2000). This ancient Greek phrase describes an organization that is both *authentic* as well as *zotikos*. The latter word means being of true significance to the people in our organizations. These are the preferred types of organizations for implementing change successfully and the people working there are able to focus their energy. They contribute to the organization's success with measurable effects in terms of quality, cost awareness, and profitability. Within these organizations it will be possible successfully to implement innovative concepts and practices as well.

Our independence

One of the world's most well-known and respected thinkers on management, Charles Handy, wrote an entire book, *The Elephant and the Flea*, about the importance of obstinate and creative individuals in our service-oriented economies (Handy, 2002). The elephant represents today's large organizations. The flea represents the individual, who works *independently* for those organizations. Handy considers these self-willing fleas as essential for initiating the changes and innovation required within large organizations. These fleas, being truly individuals, are the keys towards change and are vital to bring the elephants to dance on the rhythms of change in society. It seems to be their independent drive to change that encourages the people around them to join in.

At this point I should to say a few last words about the concept of the individual. This is meant as a genuine warning, to keep our individual feet firmly on the ground. Let us stay modest and humble about the concept of the individual. Dewey (1927) mentioned many years ago that any individual does not really think independently, but simply expresses in a unique way the thinking at that time. The ancient Greek philosopher Plato expressed his thinking in carriages and horses. The great and early twentieth-century psychiatrist Sigmund Freud expressed his theories in the mechanical terms of forces and machines. The future will show how our current thinking about the concept of the individual is merely an expression of our time.

The concept of perception

Without exception, if you work in the construction industry, in any function, you will be confronted with the aspect of ground conditions. Every project seems to get calamity, small or large, related to the ground. I thought that after such an event the geotechnical consultant was wrong, but now I know that every consultant could be right. The difference is the individual *risk perception*, related to the intrinsic uncertainty in ground conditions.

Joost Wentink (2000), former contractor and consultant, Managing Director of GeoDelft

The word *perception* has two meanings according to my dictionary. The first is that of *receiving*. The second meaning is the *result of observation*. The term 'perception' therefore goes beyond the meaning of just looking. The difference between looking and perceiving is that perception gives some kind of *sense* to the observation result. In other words, combining the two meanings of perception is to receive an observation and to give a certain *sense* to it. Perception appears to be closely related to interpretation.

Why this rather semantic start to this new section? What is the relevance of such definitions and meanings of words that are widely used in day-to-day communication? The reason is that a true understanding of the concept of perception serves as the ultimate basis for understanding risk management in general, and risk management of ground conditions in particular. Perception drives the way in which a person considers risk. It implies that risk and its management have a highly subjective character, in spite of the apparent objective statistics often linked with risk management. This is due to the major effect of the *people factor*, the high touch element it incorporates. As the ancient Greek philosopher Aristotle stated some 2000 years ago: 'You are what you see'.

Subjective perception

By looking at two examples, careful self-observation may teach us that apparent objectivity is not as objective as it seems. It can be demonstrated that our perception is rather subjective, with different people having quite different perceptions, based upon exactly the same factual data. How confusing for an engineer. We apparently have our own individual way of looking at things, seeing only part of the whole picture most of the time. Let us start simply and have a look at Figure 4.1. Which line is the longest?

When I first saw these two lines, the upper line appeared to be the longest. After a quick check with my ruler, however, I discovered that I was misled by my own perception. The two lines are in fact equal in length. This type of difference in perception, the difference between what seems (*subjective*) and what is (*objective*), is sometimes easy to resolve, as in this example. It becomes far more complicated, when this objectivity is less easy to measure. A picture of the so-called ambiguous lady, a well-known example within psychology, demonstrates

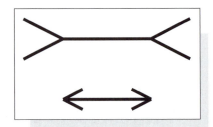

Figure 4.1: Which line is the longest? (© Blockley and Godfrey (2000) with permission of Thomas Telford Ltd).

this. The psychologist W.T. Hill, in 1914, drew a picture that has been copied in many different ways since (Turner, 1995). One version is shown in Figure 4.2.

Figure 4.2: The ambiguous lady *(© Blockley and God-frey (2000) with permission of Thomas Telford Ltd).*

If you look at Figure 4.2 for the first time, you will probably recognize either a pretty girl or an older lady. Both of these quite different perceptions originate from exactly the same drawing. The lines in the drawing are a fact. Our mental interpretation or configuration, using those lines to give a mental picture, can apparently provide different outcomes. In other words, there is a clear difference in individual perception, based upon the same facts. By our observations there are more ways to give sense to the same facts. I have used this figure a couple of times within groups of people and am still amazed by its effect.

With some practice, you can probably see both the young girl and the older lady. This demonstrates that we are able to see multiple perceptions. At least two of them, according to this example. We are able to *add* another perception, apart from our initial one. Nonetheless, these two perceptions of one individual are based on the same factual data.

The preceding statement of Joost Wentink has mentioned a similar difference in perception of ground conditions by several engineers or geologists. Based on the same factual data from ground investigations, different interpretations are likely to arise, which result in different calculations, as well as differences in engineering design, as demonstrated by a few examples in Chapter 5. Further in this chapter, we will appraise the ability of conscious *shifting of perceptions* as a key success factor for effective risk management.

Another remarkable aspect of human perception is related to numerous words that we often unwittingly use. The psychiatrist, Ron Leifer, describes this as the dialectic of *antithetic pairs* (Leifer, 1999). These are pairs of words with an opposite meaning. The word *hot* only has a meaning in connection with the word *cold*. If there is no coldness, then the meaning of hot is, in fact, empty. Any temperature within the continuum of hot and cold obtains only a meaning by its benchmarking with the words hot and cold. This meaning will probably differ for each person. An Eskimo will relate another temperature to the word cold than someone who lives in the tropics.

We will now approach the inherently fuzzy character of ground, resulting from our inherent differences in perception. Unless we quantify it using some definition, the term hard rock will have a different meaning for different professionals. The inherent subjectivity in the use of our words, as well as their intended purpose, will only surface if we communicate carefully with each other. Effective communication serves as the basis for dealing with the different perceptions we encounter daily in our social environment.

Dealing with different perceptions

Anyone we approach will have different perceptions to our own. These differences may be small and barely recognizable, but they may also be huge and act as basis for disastrous conflicts.

Dealing with a variety of perceptions is not simple when carrying out our ground-related engineering and construction activities. Frans Barends, professor at the Delft Technical University, The Netherlands, concluded in his Terzaghi Oration 2005: 'The effects of subjective individual interpretation of facts and data are underestimated.' He preceded this firm conclusion with a geohydrological example that could dramatically affect the perceived dike stability. His example concerned the interpretation of an observed porewater response, caused by changing water levels. This interpretation proved to be highly subjective, as six different geohydrological models could be applied to interpret the data. Selection of the type of model in fact proved to be a subjective choice. Extrapolation using a well calibrated, although incorrect, model may give rise to unexpected behaviour. According to Barends (2005), this is often the reason for failure and damage. In my opinion, it results from a lack of awareness about the subjectivity involved in ground engineering and construction, combined with of a lack of explicit risk management. Ho et al. (2000) warn us that an open and risk driven approach towards individual subjectivity may be well considered as a threat by individual professionals in our industry. According to them, it would be more difficult to hide behind their so-called overall expert judgement. It reflects the human nature of professionals that is also occasionally encountered in other disciplines, such as the medical discipline.

Preferably, we should not only be aware of different perceptions, but should also need to *understand* them to a certain degree. By evaluating the behaviour of chimpanzees, O'Connell (1997) attempts to understand how we develop our abilities to recognize how someone else perceives the world. She distinguishes and highlights two well known but often intermingled terms: *sympathy* and *empathy*. O'Connell defines *sympathy* as reflecting our own perception. If we consider someone to be sympathetic, we like them because of the congruence of that person with our initial perception. They are in fact checked and balanced against our own perception of the world, according to our own beliefs, intentions, and wishes. In the case of *empathy*, however, we are able to see someone else according to *their* perceptions, instead of our own. We now make a shift in perception in order to understand the beliefs, intentions and wishes of the other person. For many people, such as myself, the application of true empathy is rather difficult. We need to be aware of it and practise it, as it is the key to deal effectively with the inherent variety and differences of people, both from our own and other cultures. We will encounter these differences more and more in our increasingly global construction world. Fortunately, differences in perceptions are not always difficult and annoying, as presented by Box 4.1.

Box 4.1 Dealing with different perceptions at a holiday resort

During a diving holiday in Egypt, a few years ago, I noticed that entirely different perceptions do not automatically result into difficulties or conflicts. On the beach at a holiday resort, people of many different nationalities, with totally different cultural backgrounds and related perceptions, enjoyed their holiday together. They varied from young girls in tiny bikinis to older women dressed entirely in black. Sometimes we can simply live together with our entirely different perceptions of the world.

Additionally, in the same way that risks may reveal opportunities, different perceptions may also be extremely helpful. A practical advantage of a difference in perception is illustrated in Figure 4.3. It shows a detail from my site office at a project in Indonesia in the 1990s, just before the current era of the mobile phone. In my perception, I had to walk to another room to make a phone call. My neighbour had a much more efficient solution.

In conclusion, why is it important to be aware of all these differences in individual perception? It is because they incorporate a variety of *risk* perceptions as well. Effective risk management is not possible without being aware of the differences in individual perception. This is not new at all. Some thousand years ago it is made clear in the ancient Chinese philosophy of the Tao, which literally means 'the path'. From time to time we should return to what is known as *the zero*, in order to achieve an as much as independent and objective perception as possible (Ni, 1997). We will sometimes need this zero-point urgently when dealing with ground-related risk management.

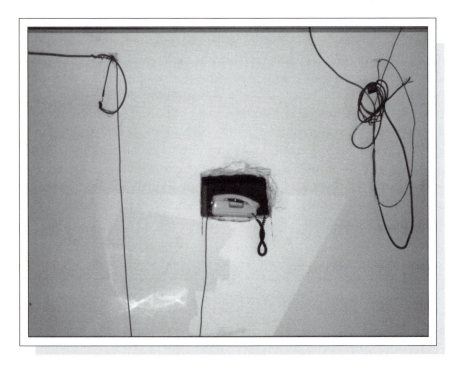

Figure 4.3: The beneficial side of a difference in perception (© photo by Martin van Staveren).

Individuals and risk perception

How may individual professionals perceive risk management? A manager of a large contractor, I call him Bob, told me how his individual employees perceived risk and its management: 'First I am asked to dig my own holes, by telling some sort of risk coordinator what can go wrong by my activities. Then I have to convince that same person what I do to prevent myself from falling in my own holes. I have to present that guy all kinds of so-called risk remediating measures'. Bob continued: 'This type of risk management is not motivating my people!' In other words, Bob's employees are fed up with such management of their risk. They feel forced to do it and do not (yet) perceive risk management as a tool that can help them rather than providing additional hassle in their daily busy construction activities. Risk management is perceived as a waste of time.

While Bob himself is dedicated to the concept of risk management, despite his struggle to find the right way to convince his employees about the benefits, any attempt to implement risk management in such an organizational setting is useless. An open discussion about the differences in risk perception, starting with the purpose and expected benefits of it for all people involved, is of paramount importance. Without a number of serious discussion sessions with Bob and his employees, their perceived differences about the purpose and benefits of risk management remain. Only after thorough communication may Bob's employees become aware that risk is simply there and that it is better to be managed, as they manage the other aspects of their activities as well. The employees may begin to realize that risk management is not a tool to highlight their weaknesses or mistakes, but intended to support them in avoiding the pitfalls already provided by reality. For Bob, his employees and the final project result, it will be much better discovering these hidden holes, to prevent their acting like sinkholes in limestone. Awareness, understanding and acceptance of the inherently different perceptions between Bob and his individual employees is a key issue to come out of this type of situation, in which an implementation of a new concept is seriously obstructed.

Complexity of individual risk perception

A lot of managers experience it: people are not only interesting, they are complex as well, particularly with regard to their individual risk perception. Smallman (1998) presents three theories of risk perception to explain how we perceive risk. Wildavsky and Dake (1990) tested each of these. The first theory, the *knowledge theory*, perceives risk as problematic because of the knowledge we have of it. I recognize this theory very well in the practice of ground-related risk management. A geohydrological engineer focuses particularly on geohydrological risks, while the environmental engineer is primarily concerned with environmental risk, and so on.

The *economic theory* of risk perception translates risk into a subjective utility. Risk is judged in view of the expected satisfaction or, more likely, dissatisfaction, which its occurrence will bring. This dissatisfaction is often expressed in money terms. Ground risk management widely uses the economic theory of risk perception and the effects of ground-related risk are often assessed in terms of costs.

The third and last key theory of risk perception, the *personality theory*, puts emphasis on the personality or character of individuals. Some of us are more risk adverse, others are more risk seeking. In my view, our personalities are influencing both the knowledge and the economic theory of risk perception. The knowledge we use to assess the cost effects of risk should be expected to be coloured by our personalities. A risk adverse type of person is likely to assess a particular risk effect more seriously than a more risk prone character.

The combination of these three risk theories explains the complex character of individual risk perceptions. In addition, risk perception is affected by a number of social and cultural factors. Rohrmann (1998) adds this social context as a third dimension to his two proposed dimensions of the individual context and the character of the considered risk. In his model these three aspects determine any individual risk perception, that is herewith inherently subjective. Figure 4.4 presents the three pillars below individual risk perceptions.

The three dimensions, as proposed by Rohrmann (1998), are obviously highly interdependent. A social–cultural aspect, such as the degree of media attention, has different effects on different personalities. Furthermore, any inherently biased individual assesses the seriousness of a certain risk in a different way, with a different outcome. Unfortunately, even if we are experts in our disciplines, it does not delete our subjectivity. Research by Slovic et al. (1982) indicates that the judgement of experts is as prone to bias as that of lay people.

Awareness of the uniqueness of risk perception is a

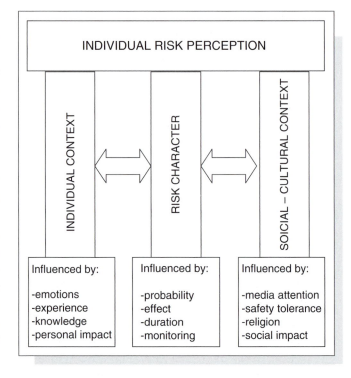

Figure 4.4: Three pillars below our individual risk perception.

major step in the application of (ground) risk management. Within this context, effective risk management can be summarized as managing different individual risk perceptions in an effective and socially accepted way and against reasonable costs. Additional complexity arises, however, when we recognize individual risk perceptions as being not static. They are subject to change over time and distance.

Change of individual risk perception with time

Let us consider a construction project, for instance a road project in its design phase. We are asked, being an engineer or a manager, to support the design team during a risk identification and classification session. Our individual and unique risk perception of that particular road project, at that very moment, depends on a lot of factors. Without assuming completeness, some of them are:

- Our *experience*: do we have good or bad experience with a similar project, in similar geological areas, or do we not yet have any experience at all with road projects?

- Our *knowledge*: what did we learn during our BSc or MSc courses about this type of project?

- Our *personality*: do we generally have an optimistic character, or are we more of a realist, or even a pessimist?

- Our *mood* of the moment: do we feel good, or do we feel tired and worried, for instance because of an illness of one of our family members?

- Our *company* at that moment: is our company struggling for survival because of very serious price competition and is any design work highly wanted, at almost any price? Or is our company leading in its market and well known for its innovative approach of road projects?

We may recognize some aspects of the knowledge, economy and personality theories of individual risk perception in these factors, while realizing that most of these factors are far from static. They will change over time, such as our mood may turn to quite pleasant, after a good-news phone call. Additionally, over the years our knowledge may have grown substantially by taking a postgraduate risk management course. Our company may change as well, for instance because of a merger, new management or the launch of an innovative service. Even our personality, our unique set of personal characteristics, will have some degree of change over time, simply as a reaction to all these changes around us.

This brings us to one conclusion: our individual risk perception is *not* a constant factor. The same applies for all the other individuals, our project team members and clients, government representatives and the public living around the project area. Their risk perceptions will also change, due to the constant flow of new information and experiences they undergo. All of these people absorb this information, from moment to moment, both in an aware and unaware manner. As the ancient Greek philosopher Heraclitus of Ephesus mentioned some 2500 years ago: *Pantha Rei* – everything flows and changes continuously. I admit directly this it is a very complicating factor in risk management as we need to manage something that is changing constantly, often in a quite unexpected way as well.

Change of individual risk perception with distance

As with time, risk perception also changes with *distance* to the object of perception, which provides another rather complicating factor. In this respect, the individual degree of involvement with a risk should be regarded as some sort of distance as well. At a greater distance, we see obviously less detail and more of the bigger picture. We can see the wood and not only some trees. Additionally, we are less directly involved as well.

A rock engineer will see a lot of subtle *rock material* details at a small distance to a rock outcrop. By using his magnifying glass, the engineer is even able to distinguish rock minerals and to classify the rock type accordingly. To understand the *rock mass* properties as well, the engineer has to step backward to consider the discontinuity pattern. Aerial photographs can be used to arrive at a much more larger distance and to compare the rock outcrop with another one in the neighbourhood. These differences in viewpoint to the rock outcrop help to assess for instance a rock-fall risk.

The *effects* of the rock fall on an adjacent road may be quite differently assessed by the rock engineer, who is living far way from that site, compared with a local farmer, who has to pass the road a couple of times a day. The individual involvement of the expert is high with regard to the technical part of the rock-fall risk, but rather low in view of its impact. This simplified example highlights the considerable confusion and misunderstanding that may arise when experts and lay people, with different degrees of individual involvement, have to judge the same risk. Their resulting different risk perceptions are importantly caused by their different degrees of involvement or distance.

Therefore, to arrive at an effective and acceptable risk assessment of the expected rock slope behaviour, we must be aware the effects of these difference on the resulting risk classification. While considering our own risk perception in a particular situation, it is highly advised to consider our own actual distance at that very moment, both literally to the object at risk as well as to our degree of technical involvement. It may be fascinating to experience the differences in perceptions when we change our own distances for a while. Obviously, this can be done either in reality or by our own imagination. It is just a small exercise that may help us to stay modest about our own individual and inherent subjective perception of (ground-related) risk.

How individuals can contribute

Awareness of our individuality with its characteristics may surprisingly support making a difference by *ourselves*. Within the topic of this book, our main objective may be implementing ground-related risk management in construction projects. As we previously concluded, the application of explicit risk management can act as a catalyst to the urgently required construction industry transformations. How can *we*, as individuals, actually bring about the change? From experience I have selected six principles (Figure 4.5), which may be of support during the occasionally tedious implementation processes of risk management.

The remaining part of this section discusses each of the principles in Figure 4.5.

Risk awareness

It all starts with risk awareness when it comes to risk management. Each professional, in some way involved in risk management, should at least have some

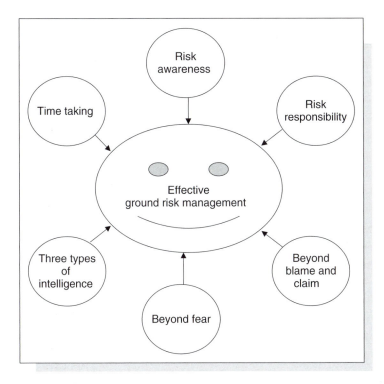

Figure 4.5: The individual contribution to effective ground risk management.

basic understanding of the fundamentals of individual risk awareness. These were extensively explored earlier and can be summarized by the knowledge, economy and personality theories of individual risk perception, which included the inherent differences in risk perception between people. Furthermore, complexity of risk perceptions, caused by its changes over time and distance, needs to be acknowledged. This type of individual risk awareness is a prerequisite for a constructive attitude to effective risk management in our projects.

Risk responsibility

There seems to be a tendency to outsourcing of individual responsibility. One example is the development an alcohol-sensitive car key. The key prevents starting of the car when it smells alcohol and the driver has to look for another way to come home. This is, in fact, an outsourcing of individual responsibility for safe driving. The *car* is going to make the decision about safe driving, instead of the individual driver. This example may soon become reality, because a number of car manufacturers are considering production of this type of key.

Let us recall the collapse of part of the Terminal E2 of Charles de Gaulle Airport in Paris. In this 750 million euros project some 400 (!) construction firms, from all over Europe, were involved. How is it possible to manage risk in this sort of complex project when individuals are not taking their risk responsibility, but are sourcing it out? The resulting fragmented and fuzzy responsibilities are a serious threat to the project safety and success. Critical risk fragments may slip through the loopholes of the quality system, because the people factor of individual responsibility is very difficult, if not impossible, to catch in procedures. Despite its possible short-term drawbacks, a risk-responsible attitude of professional individuals will finally be in the interest of their projects, clients, organizations and themselves.

Blockley and Godfrey (2000) have considered a number of existing professional codes of conduct for the construction industry. They have derived twelve key-issues, shown in Box 4.2.

Box 4.2   What responsible construction players should (after Blockley and Godfrey (2000), with permission of Thomas Telford Ltd.)

Twelve key issues for responsible construction players are:

1   Be clear about what their own values are

2   Be informed about the project in which they are involved

3   Think about the consequences of what they do and in particular try to anticipate the unwanted and unintended consequences

4   Be up-to-date in professional skills

5   Act professionally only in their areas of competence

6   Keep health and safety and public welfare paramount

7   Communicate openly with the public about technological developments

8   Be honest

9   Disclose circumstances where there may be a conflict of interest

10   Neither offer nor accept bribes

11   Treat all others fairly in respect of race, religion, sex, age, ethnic background or disability

12   Help colleagues promote growth of skills and competence

Paine et al. (2005) have performed a similar exercise. They have studied and compared five recognized sets of guidelines for multinationals and were surprised about the degree of similarity between them. Any differences between these codes of conduct were complementary rather than conflicting. There is apparently wide international agreement of responsible behaviour in global business. However, according to Paine et al. (2005): 'Like any tool, a code of conduct can be used well or poorly – or left on the shelf to be admired or to rust.' This is up to us!

Beyond blame and claim

Increasing blame and claim behaviour appears to be another trend. Smokers blame the tobacco industry because they got lung cancer and hamburger addicts blame fast-food restaurants, as they grew too fat. The construction industry has also a claim tradition. For a significant number of contractors, claiming seems the only way to make some profit and a substantial part of the failure costs can be related to these claims.

According to Block (2002), feeling much more convenient by blaming other people, rather than ourselves, is a normal human condition. In addition, Imai (1986) highlights a unifying characteristic of a lot of problems: the people or organizations that create problems are often not directly inconvenienced by it. The same applies for risk, which can be considered as a specific type of problem. A risk that affects somebody else may well return as a claim, as presented by an example in Box 4.3.

Box 4.3   A blame and claim example

Let us consider a geotechnical engineer who has designed a very safe piled foundation. The piling contractor encounters the risk of very high blow counts during his piling activities. The over-conservative pile design of the geotechnical engineer causes a number of broken piles. In addition, the vibrations of the heavy piling operation cause serious cracks in the walls of some old and sensitive buildings, located close to the project site. The owners of these monuments claim from the contractor or the contractor's client to pay the costs of the repair of their buildings. In response, the contractor or client claims from the engineer for providing an over-conservative foundation design. None of the parties involved is *directly* affected by their activities.

We are thus sensitive to problems and risks caused by other people. In return, we are often rather insensitive to the problems and risks that we cause to others. The possibility of being blamed and claimed by someone else, however, keeps us awake and aware of our own responsibilities to other people and organizations.

Steven Covey is the well-known management guru who created the seven, and later eight, habits of highly effective people. He relates blaming other people to three stages of dependency that we can reach as human beings (Covey, 1992):

- Dependence

- Independence

- Interdependence.

The first stage is that of *dependency*, because we all start our little lives in total dependency on our parents. Therefore, if something not so nice happens to me, then it is not my responsibility because I am totally dependent. An external factor or another person causes it. In this stage I blame anybody but myself.

The second stage is that of *independence*. In this stage we are able to blame ourselves when something goes wrong, caused by our own activities. We take our responsibility and realize the effects of our choices and doing or not doing.

Covey's final stage of dependency is what he refers to as *interdependence*. It is not about an isolated you or me anymore, it is about us, about we. It is an awareness of everybody being in some way dependent upon the other. Therefore, it seems a wise *choice* to cooperate to create win-win situations. An effective application of (ground) risk management requires this type of interdependence awareness, which is, I agree, a rather ambitious statement. If we simply try to start raising awareness of the interdependency-concept in our projects, we actually already start to move beyond blaming and claiming.

Beyond fear

While taking our individual responsibilities and trying to reduce construction's blame and claim culture, we have to look another very human condition straight in the eyes: *fear*. This human emotion prevents us from doing stupid things. During our evolution, fear allowed us to survive as a species and, being normal people, we all encounter fear from time to time.

Fear can be a very strong emotion because it is a response to some kind of threat, a real one or an apparent one. The problem part of this emotion is preventing rational thinking. Emotions create *impulses* to act without the interference of thinking, which is reflected in the meaning of the term *emotion*. It is derived from the Latin word *movere*, which means moving. The letter *e* in *emotion* stands for moving away (Goleman, 1996). If we encounter a fearful situation, the immediate response is flight or fight. Normally, if the situation is not extreme, rational thinking soon takes over control and we continue to behave as decent people do.

Fear appeal can be defined as attracting business by triggering the fear of people (van Oirschot, 2003). Occasionally, when reading the newspaper or watching the television news, I get the impression that we are living in some sort of *fear factory*. A lot of fear appears to be blown up by several media and, apparently, we like to be fed with fear-food. One example of a massive business created by fear appeal was demonstrated just before we entered the new millennium by the so-called *millennium-bug*. The ICT industry succeeded in convincing all other industries to update their computer software to avoid disasters when the new year with the magic figures 2000 started. Worldwide, huge sums of money have been spent and, for sure, we will never know the effects in case we did spend less to resolve this collective fear.

Regarding risk management, why are fear and its appeal reasons for concern? Fear and fear appeal are of massive concern from a business perspective, because fear-driven behaviour results in over-investing in risk management, without receiving the expected returns. We are buying some sort of emotional quietness at a too high price. Our rational factors for deciding on investments are at least seriously affected, and possibly even overruled, by these fear factors of emotional origin. If there is an apparent overkill in ground risk management during a construction project, initiated by implicit fear of one of its decision-makers, such risk management would probably be experienced as highly disappointing by the majority of professionals involved. In their next project, people with this type of risk management experience are likely to under-invest in or even avoid risk management.

Each risk that we manage successfully consequently does not occur in its full extent. We will therefore never know exactly what we saved by it, because the adverse effects cannot be experienced, measured and benchmarked. This inherently hypothetical character is one of the key paradoxes of risk management. I have to admit that I might have created some fear appeal, by previously presenting the figures of failure costs in construction. I would regret, however, when risk management is only applied because of its fear appeal. In this situation, emotional flight and fight behaviour rules instead of rational thinking. It would put too much weight on the negative problem side of risk, while neglecting the positive opportunity side.

I do not intend to neglect human emotions, nor do I underestimate them. They give a lot of colour to our daily (working) lives. However, risk management in the construction industry, and ground risk management in particular, should not be based upon fear. This situation will create an unfavourable and defensive attitude and prevents innovation and effectuating cost-effective opportunities. Too often, I encounter good-willing professionals who do not dare to innovate just because they fear uncertain outcomes. Of course, they do not explicitly reveal their sources of fear, which is not done as a construction professional. Instead, these professionals use all their creativity to argue why not to innovate. Nevertheless, they have a risk-averse attitude. I wish they would use their creativity and energy to support innovation instead of fighting *against* it.

Expressing fear clearly is a taboo in many organizations. However, often it may be far more effective to recognize fear, to be aware of it and to use it in a positive way than to neglect fear.

This individual fear awareness and management needs to be considered as being an integral part of risk management. We occasionally have to look these fears straight in the eyes, in order to manage them. The same applies for risk. According to Blockley and Godfrey (2000), most issues carrying a risk are known to at least one member, one individual, of the team. These risks are not always raised, because to do so, we have to enter our zones of discomfort. A number of reasons may prevent this move, as presented in Box 4.4.

Box 4.4 Zones of discomfort (after Blockley and Godfrey (2000), with permission of Thomas Telford Ltd.)

Some reasons for not bringing potentially risky issues into the open:

- The possibility of a future claim and a potentially better case for that claim if the problem gets worse

- Self-protection

- A culture of blaming, rather than addressing real issues

- A tendency to 'shoot the messenger'

- Competitive pressures between teams, e.g. sales teams, design teams, procurement team, construction team

- A sense of loyalty to one's client, colleagues, teams or employers

- People just do not want to jeopardize personal relationships by opening them up

---

Fear can be recognized as being the dominating factor in the zones of discomfort. It prevents us from entering these zones, in spite of its potential benefits. Box 4.5 presents a few personal experiences to demonstrate that entering these zones of discomfort may pay off, both in our personal and our business lives.

Box 4.5   A personal note about zones of discomfort

Due to expected discomfort and fear, there is often a small and fragile boundary between acting and not acting. My diving experiences form one example. I got an opportunity to learn scuba diving on my first overseas project in Saudi Arabia. Initially, I was seriously worried about diving. I could not even snorkel in a decent way. The enthusiasm of my colleagues made me enter my zone of discomfort. I got my diving certificate and had great times. The Red Sea proved to be a diver's paradise, even for the novice diver.

I had to enter similar zones of discomfort during a number of geotechnical assignments abroad. Of course, I was worried while starting these adventures. Can I achieve the objectives? It was sometimes tough, indeed. However, all of these projects turned to be interesting experiences in which I was able to make some difference. This was only possible by entering my zones of discomfort.

Entering our zones of discomfort appears occasionally to be necessary. Fear should be challenged, of course in some balanced way. In my opinion this attitude is required to make a difference, for instance by implementing new concepts, such as risk management, in often change-adverse environments. In this respect, I recall Imai (1986) who stated: 'There will be no progress if you keep on doing things exactly the same way all the time'. Sometimes we even need to move beyond fear to make progress.

Three types of intelligence

While still subject to serious debate, since the 1900s, intelligence is often assessed by the so-called Intelligence Quotient (IQ). This IQ aims to classify particularly our rational or logical sort of intelligence. It mainly concerns our *thinking*, which we doubtlessly need to solve many technical challenges within every construction project. We use this intelligence for a wide range of activities, such as providing cost calculations for tendering and complicated finite element calculations for all sorts of geotechnical design. Since the 1960s, it became apparent that IQ does not cover the entire spectrum of intelligence. The psychologist Daniel Goleman has provided a real break-through by introducing the Emotional Quotient (EQ), a measure for emotional intelligence. Our EQ concerns particularly *feeling*, rather than thinking (Goleman, 1996). We need it for handling the people factor in our projects.

Contrary to IQ, which is more or less fixed, we seem to be able to cultivate EQ and even to bring it to a higher level. A vital characteristic, because this may improve also our abilities to apply our IQ. So long as we do not know how we feel or what we feel, our emotions will largely rule our attitude and behaviour. This will blur and even decrease our capabilities of rational thinking. According to Goleman (1996), self-awareness or the ability to recognize our own feelings, is the key to improve our EQ. One of the ways to increase our self-awareness is meditation, as practised for thousands of years within many cultures worldwide.

At the end of the 1990s, neurological research confirmed that we may even have a third type of intelligence: the Spiritual Quotient (SQ). It is about *being* and concerns who we are (Zohar and Marshall, 2004). In this context, the term 'spiritual' does not have any specific religious meaning. It originates from the Latin word *spiritus*, which means something that brings life or vitality to a system. Therefore, spiritual should be perceived in relation to *meaning*, *values* and *fundamental objectives*.

Being a ground engineer, spiritual intelligence sounds a little bit floating above the ground to me and I prefer to speak of *inspiring intelligence*. This can be seen as the capability to ask ourselves rather existential questions like: What do I want to reach by finalizing this construction project? Why do I have the job I have? What is my purpose of life? Additionally, SQ may also provide us the drive to inspire and motivate other people around us to change and innovate with us. Let us bring the spirit in the ground!

In the opinion of Zohar and Marshall (2004), the conventional IQ-type of intelligence operates within the boundaries of a sys-tem, which can be con-sidered as the hard systems approach with fixed bound-aries. The EQ-type of intelli-gence is more adaptive than the IQ-related intelligence. It serves to play the game within a social context by the conscious or perhaps unconscious application of the social rules of any group

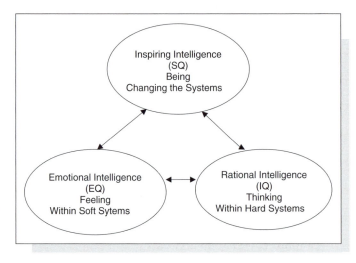

Figure 4.6: Three types of intelligence.

of people. Still, this game is played between boundaries of a soft system. The SQ-type of intelligence may *change and redefine* these soft systems boundaries. This type of intelligence facilitates transformation, how it *is* today is transformed to how it may *become* tomorrow. It gives the power to leave behind worn-out old paradigms and to create effective new ones (Zohar and Marshall, 2004). Figure 4.6 shows the relationship between the three distinguished sorts of intelligence.

To conclude, we obviously need all three types of intelligence in a well-balanced way to succeed in our construction projects.

Time taking

The last principle for an individual contribution to effective ground risk manage-ment is probably rather unexpected. I may be starting to move on slippery ice, according to an old Dutch saying. However, I think we should move like this, once in a while, to take some distance and to reflect on our common patterns of thinking and behaviour.

Often it seems we are performing some kind of rat race, which is both chal-lenging and exhausting as well. By just rushing forward, we might miss essential insights and opportunities. An old Chinese saying teaches us: If you are in a hurry, sit down!

My last recommendation is therefore to take sufficient time for reflection. Not by taking enough sleep and days off, but just by few short periods of reflection during the working day. It is a simply a matter of sitting down or standing at rest, being quite and aware of our breathing and, most important, doing nothing for a few minutes. This is, in fact, the easiest way of practising some sort of meditation (Witten and Tulku, 1998). While meditation has been applied for thousands of years, in many parts of the world, it slipped away from the western life-style. I recognize, however, a comeback, for instance reflected by the many books about this topic that have been published in recent years. For an increasing number of professionals, daily meditation is a fascinating and rewarding experience. It is amazing how relaxed and clear-minded you may feel, just after these small breaks a few times a day. Not to forget, as suggested by Goleman (1996), meditation strengthens the individual self-awareness as well. This will support us to act as individual change agents.

The individual change agent

By now we are equipped with a number of mental attitudes and techniques that prepare us to start our journey as being a change agent. The role of the dedicated change agent is paramount in the abundantly available management literature about change management and organizational transformations. For instance, Roobeek et al. (1998) stress the importance of *strategic dialogues*, in open network-type of organizations, in order to realize real transformations. Many different professional individuals should be involved, not just executives and management. Gratton (2004) advocates the development of what she calls *democratic organizations*, where employees perceive and encounter each other as responsible and mature professionals. Change agents are, however, required to initiate and realize these types of ambitious organizations.

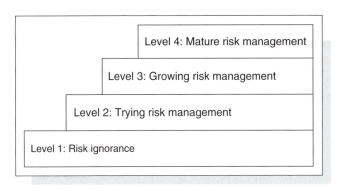

With regard to risk management in project organizations, the maturity concept makes sense as well. Edwards and Bowen (2005) draw upon the work of Hillson (2002) with regard to the concept of risk-maturity. They define four ascending grades to *risk maturity* for organizations. These levels do not only apply to organizations, they may very well be applicable to individuals as well (Figure 4.7).

Figure 4.7: Ascending grades to individual risk-maturity.

By now we should have left Level 1 forever. The six principles for an individual contribution to effective ground risk management are dedicated to rising further on the steps of individual risk maturity.

This chapter started with a phrase of the ancient Chinese philosopher Confucius, who lived about 2500 years ago. Times have changed since, however, some statements appear to be timeless, as demonstrated by a phrase of one of today's most successful entrepreneurs, Steve Jobs, founder and CEO of Apple Computers. He concluded a commencement address for graduates of Stanford University with the words:

> Your work is going to fill a large part of your life, and the only way to be truly satisfied is to do what you believe is great work. And the only way to do great work is to love what you do (Jobs, 2005).

Summary

This chapter has explored the relationship between the individual and risk. The concept of the individual has revealed discontinuities not to be limited to rock masses. We may have inner discontinuities as well, when we struggle with authenticity and the gap between our professional and personal selves. These discontinuities may adversely affect our attempts to bring about the necessary changes in our project organizations.

The concept of perception, together with a number of examples, have taught us the inherent subjectivity of perception. Each individual has a different perception and we need some basic empathy to cooperate effectively with this fact in our daily working lives. Particularly for effective risk management, we need to appraise and judge these differences in perception, while remembering the inevitable bias of both expert and lay people. This complexity of risk management even increases due to the changes in risk perceptions over time and with distance.

There is good news as well; individuals are able to play a key role in the implementation of change initiatives, like the application of (ground) risk management in project organizations. Six key factors for individual professionals have been distinguished and discussed thoroughly: risk awareness, risk responsibility, beyond blame and claim, beyond fear, three types of intelligence and time taking. Their awareness and application may highly support the individual change agent to make his or her difference. This has prepared us to enter the project team.

5 Teams and risk

Introduction

Teams are of paramount importance for effective ground risk management. However, they add complexity to the already complicated behaviour of individuals. Within teams, individuals tend to adapt their attitude and actions, which may create positive as well as negative effects. Insight into the behaviour of teams and their approaches to risk will help to implement ground-related risk management in our projects.

We start with appraising the concept of the team. Groups and teams need to be distinguished and we have to go through a number of inevitable phases when creating a team out of a group of people. Particularly with regard to uncertainty and risk, the concept of team culture further facilitates our understanding of teams and their dynamics.

Our day-to-day practices teach us that communication between people is not that simple. Better formulated, communication between people is inherently complicated. Teams and their risk communication, therefore, get particular attention in this chapter.

Finally, three types of team demonstrate how teams can contribute to the introduction and application of ground risk management. For the moment, the client is not yet considered as a team member, because their interests differ importantly from those of the contractor or the engineer. The client in our team is not yet that common in our industry, although innovative ways of cooperation with clients, such as partnering, have been started. Nevertheless, the client and risk will be described separately in the next chapter. As usual, this chapter closes with a summary.

The concept of the team

From groups towards teams

Aside from of how clever we are, today's construction is too complicated to work on our own. We need more people to be effective and efficient. It is therefore no coincidence that the majority of people live in societies where group interests are considered as more important than individual interests (Hoecklin, 1995; Hofstede, 1984). In these societies the groups form the basis of the identity of their members.

More than one person being together can already be called a group. In any group there is some sort of interaction between the individual people. They influence each other, both consciously and unconsciously. This social component complicates proper understanding of the attitude and behaviour of individuals in groups.

What makes a group of people a team? What is a team anyway? Teams go beyond summing up the individuals. According to Katzenbach and Smith (1994) teams can be defined as a small number of people with complementary skills, who are committed to a common goal and approach, for which they hold themselves mutually accountable. There is a clear dependence between our team members and us. We need each other to act more effectively than if we work on our own. It is the magic of one plus one being more than two.

Abundant management literature considers effective teams as a key factor to the improvement of all sorts of business processes. However, we have to invest in order to arrive at effective teams. Without dedicated attention, teams stay merely groups of individuals and one plus one may be even less than two.

If these teams are widely considered as that import-ant, then how do we actu-ally form them effectively? The process of creating a team follows four particular phases. Before any team is

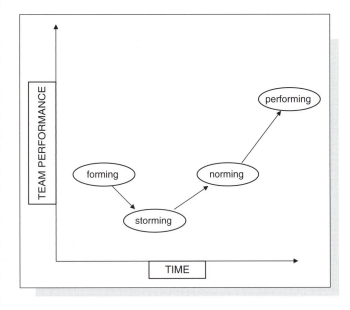

Figure 5.1: Team dynamics.

able to *perform* effectively, there are phases of *forming*, *storming* and *norming*, res-ulting from the interaction of individuals in groups. As shown in Figure 5.1, the team performance clearly changes over time as a result of these different phases.

After the forming phase of the group, in which people are normally still rather polite to each other, a storming phase should be anticipated. Politeness stops and is replaced by hidden or even openly performed arguments and clashes between team members about several topics, such as who is in charge. This lack of quietness in the team may have a rather *storming* character. We should not expect really effective team performance in this phase and if the team stays in this phase, we have a serious problem. At a certain moment, the storm will die down and the positions of each team member are settled and accepted. A certain team culture or *norming* has been reached. A mutual and shared agreement about how the team members work which each other has been developed. In the ideal team, these items are made explicit, for instance by a written *code of team conduct*.

These phases of team dynamics take time. Teams should therefore start in the earliest possible project phase, with ready available resources of time and budget, to reserve the needed time for these unavoidable team-forming phases. Learning to know each other is a prerequisite for the individuals in the team. In particular, if the team is involved in risk management, its members need to be able to build awareness of their inherently different (risk) perceptions.

Effective teamwork in the performing phase builds trust, improves communication and develops interdependence between employees, as introduced before by the work of Covey (1992). According to Oakland (1993), independence plus investment in Time, Energy and Resources results in InTERdependence. This goes well beyond one-dimensional dependence of one individual to another. Figure 5.2 illustrates the route from independence to interdependence.

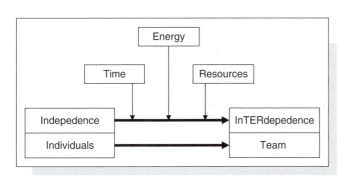

Figure 5.2: From independence to interdependence.

To conclude, distinct aspects, such as individual roles within teams, its dynamics and development stages, and team leadership should be considered in order to establish effective teams (Oakland, 1993). Furthermore, teams need to understand the nature of their problems or, more positively stated, team challenges. They need also easy access to any tools for their teamwork, like groupware-type software. In addition, teams should understand how to use their knowledge and information effectively (Uhlfelder, 2000). Consequently, it is no piece of cake to realize and run effective teams. Besides the paramount importance of the individual attitude and behaviour of team members, the team's success depends largely on one other aspect, which is *team culture*.

Teams and culture

Like organizations, smaller entities such as teams also develop some sort of typical behaviour or culture. The concept of *team culture* is rather fuzzy and as difficult to change as organizational culture once it has been developed. Team culture typically belongs to the soft system or human touch part of construction. When dealing with teams, however, the concept of organizational culture provides considerable insight into the attitudes and behaviour of individuals in groups of people.

As we have discussed several times before, at least some change in many organizations seems inevitable to survive serious competition and to run successful projects. However, the failure rate for organizational change initiatives is dramatically high. About 75 per cent of the major change initiatives failed entirely or created problems, which threatened even the survival of entire organizations. According to numerous studies, the most cited reason for failure of organizational change is the neglect of the organization's *culture*. In other words, failure to change the organizational culture will result in failure of the organizational change (Cameron and Quinn, 1998). According to Johnson (1992), a main reason is inadequate attention to the inherent links between strategy, culture and organizational processes of change. As an additional cause, managing cultural change will always be a painful process, with often strong resistance from the individual people involved (Schein, 1984).

In my view, there are few terms in business which have been used and misused as widely as the concept of organizational culture, since its introduction by Peters and Waterman (1982) in their management classic *In Search of Excellence*. Culture change appears to be some sort of panacea, widely used by management consultants, to create more effective teams, organizations or even entire industries. By the way, we also encounter the term 'culture' in a lot of the reform programmes for the construction industry.

Is it possible to define a fuzzy term such as culture? It can be described as a set of shared team values. It is about 'how things are around here', according to Cameron and Quinn (1998). Edgar Schein (1984), one of the world's most recognized experts on organizational culture, distinguishes three levels of organizational culture, as presented in Figure 5.3.

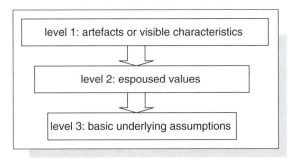

Figure 5.3: Three levels of organizational culture.

The first level contains the artefacts or visible characteristics of any culture. All team members wearing ties and white shirts is such an artefact. The second level of culture includes what Schein (1984) calls espoused values. This level is about rhetoric or what people are *saying*. For instance, the people in our project team may *say* that they are very innovative and open towards all kind of ground-related risk management applications. Unfortunately, this saying does not automatically imply these team members *act* as they say. Finally, Schein's third level of culture consists of basic underlying assumptions, unwritten rules about what people really *think* (Schein, 1984). Possibly, our team is not innovative at all because there is a serious fear of making mistakes. There might be a blame and claim sort of attitude and behaviour. In reality, our team members may be not at all dedicated to applying innovative ground risk management principles in our project.

Despite its fuzziness, several types of culture can be distinguished, each with their own features. Also, culture change can be stimulated, or even managed in some way, by changing one or more elements in Schein's three levels of organizational culture. The level one and two changes are most concrete to handle. If we are able to do it correctly, change on the deepest and most difficult to reach third level of organizational culture will follow.

The abundantly available management literature provides many frameworks for the characterization and change of organizational culture. These may be very helpful during our struggle to embed the right culture in our teams. The following combination of three different and complementing approaches proved to work well in a rather easy way. Daft (1998) has described the relationship between the external environment, strategy and organizational culture. Mintzberg et al. (1998) have linked organizational culture to seven typical contexts of organizations. Finally, Cameron and Quinn (1998) have developed an easy to handle instrument for the assessment of the current and desired organizational culture, which is based on their competing values framework. Discovering these approaches more in-depth will move us beyond the scope of this book. The references may assist you further.

Obviously, the approach of establishing the team culture depends on the type of team and its goals. *New* teams may be formed at the start of a new phase of a construction project. This raises interesting opportunities to create the most effective culture right from the beginning. A major advantage of a new team is the new mixture of people. Sometimes they even come from different organizations. The culture of a new team can more or less be made fit-for-purpose for the conditions and requirements of the particular construction project.

Existing teams have already established their own culture, either in an aware or unaware manner. Their culture may be still suitable, in view of the team objectives. However, less effective or even destructive team cultures are no exception, particularly when conditions, requirements and objectives have been changed since the start of the team. These cases call for a change of team culture, to establish stable ground for another relevant aspect of teams and risk management: risk communication.

Teams and risk communication

Risk communication

The right culture will readily facilitate another prerequisite for effective risk management by teams: risk communication. When working in a team, awareness of the differences or *ambiguity* in individual risk perception, between ourselves and our team members, is not sufficient. We have to *communicate* these differences as well, preferably in an effective way. Dibb et al. (1997) describe communication as a sharing of meaning through the transmission of information. By using the simple and well-known technical metaphor, we need a source, a medium of transmission and a sender for our communication. Feedback is the response of the receiver to a message, returned to the sender. Noise in between the sender and receiver will result in some sort of distortion of the sent information that may result in misunderstanding. Figure 5.4 presents this basic communication model.

Abundant authors, as highlighted by Edwards and Bowen (2005), present numerous and mainly more complex communication models. Aside from these models, their conclusion is firm: the effectiveness of the project team is directly related to the effectiveness of the communication between the team members.

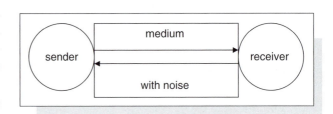

Figure 5.4: A basic communication model.

Communication between people is often experienced as complicated, also by myself, despite the enormous supply of courses and seminars aiming to improve our communication skills. A lot of our daily little and larger confusion is widely perceived as caused by misunderstanding of each other, due to insufficient or ineffective communication.

If normal communication is already this complicated, then it will become even more difficult to communicate effectively about ground-related uncertainty and risk, with its unavoidable randomness, fuzziness and incompleteness of information. Team members, often of different disciplines, with different experiences, backgrounds, personalities and interest, need to arrive at a shared understanding about their different risk perceptions. We may consider it a small miracle if we are capable of realizing this mutual understanding of risk and its management within the project team.

By building on the above definition of communication, *risk* communication can be defined as a sharing of meaning through the transmission of *risk* information. Consequently, the effectiveness of the project team with regard to its risk management can be directly related to the effectiveness of the risk communication between the team members. In addition, effective risk communication between the team and its stakeholders will be of utmost importance. A major complication arises from the fact that it is not only a matter of communicating *knowledge* about risk (Edwards and Bowen, 2005). Research by Heath, Seshadri and Lee (1998) points to aspects of *trust*, *openness* and *involvement* as key factors for successful risk communication. Which is where we arrive at the distinction between risk content and risk context in risk communication.

Risk content and risk context

Who does not want to make the miracle happen in their team: a shared understanding of risk? This shared understanding implies effective communication, for which two elements of the so-called *social–cognitive configuration theory* should be acknowledged. Watzlawick et al. (1967) describe these elements as:

1   The *content* of communication

2   The *context* of communication.

The combination of the content and context provides the *meaning* of any communicated issue. Our understanding of communication depends entirely on the content and context of what is said. This applies to our usual communication and is applicable to communication of risk as well.

The *content* includes factual information. It concerns *what* and we use our rational thinking or IQ (Intelligence Quotient) for understanding the content of the communication. It implies the cognitive part of the communicated item. The *context* of communication is about *who* and *how*. It is by far less straightforward, because of the social construction of the context. This people factor provides an additional meaning of the communicated aspect. We may relate the context to the EQ (Emotional Quotient) of the team and its individual members. The context interprets the factual content in a social framework. The ruling team culture will markedly affect the interpretation of the communicated issue, as well as the resulting team behaviour. Watzlawick et al. (1967) describe the effect of communication on behaviour by *pragmatic communication*. In their view, attitude and behaviour are integral parts of communication, like the language of the communication. Or, in the words of Berlo (1960), 'meanings are in people'.

Risk communication is, therefore, by definition, emotionally loaded, which always includes a degree of uncertainty and recalls feelings of fear with most people. According to Arvai et al. (2001), this situation rules because the core values of people are directly connected to risk. These values about what really matter to people need to be made explicit to be able to decide upon any appropriate and acceptable risk management actions. Other authors, such as Mearns et al.

(2001), support the importance of the context with regard to risk communication. As engineers, we have some natural tendency to focus on content and facts. Particularly when dealing with risk, it is not only important *what* we say, but also *how* we say it. Let us consider one example in Box 5.1, which aims to illustrate the impact of this rather complicated social–cognitive theory in practice.

Box 5.1   Soil liquefaction in a risk content and risk context

Liquefaction of water-saturated and loosely packed sand layers is a common source of risk to structures in earthquake-prone areas. Due to liquefaction, entire buildings may topple, as for instance the Japanese experienced during the Kobe earthquake some years ago.

Let us consider a design team. The team is involved in foundation design for liquid petroleum gas (LPG) storage tanks in an earthquake-sensitive area. Damage to those storage tanks may cause a disastrous situation and is considered as a major risk.

The client communicates this risk to the design team and requests an earthquake-resistant design. She provides the design team with recently performed and high-quality ground investigation data. The data set includes boring logs and standard penetration test (SPT) results, which are typically factual data widely used for foundation design in areas with liquefaction risks.

The factual data can be considered as supporting the *content* of the communicated liquefaction risk. The design team will interpret these data according to its prevailing team culture. They will create their own *context* that drives the team behaviour. This particular design team does not feel comfortable at all with the boring logs and SPT data. One of the team members once had a bad experience with apparently reliable SPTs. The design team is used to another type of test for the assessment of the soil's liquefaction potential. In their opinion, they truly need cone penetration tests (CPTs), despite the fact that another design team, with other experiences, would be fine with the ready available soil data. Because of the unique *context* of the liquefaction risk, the design team decides to put aside the available ground information, in spite of its good quality and sufficient level of detail. As a consequence of their attitude and behaviour, the design team requires the client to provide additional site investigation, to be performed by CPTs.

The team culture and its prevailing atmosphere may also determine *how* the team will communicate this risk message to their client. If they are already behind schedule, stress, frustration, and anger are likely to affect or even dominate their communication. They may create a rather negative *context* in their risk communication to the client. This may even overrule the *content* of their message and have an adverse effect on the relationship with the client.

She may be upset about the way this design team gives her feedback on her liquefaction risk message. In her perception, with over 20 years of experience, the data she delivered have always been considered as adequate by other design teams.

If the design team is in better and more professional control, they will be able to communicate less emotionally and more rationally to the client. They will be able to explain their adverse SPT experiences and inform her about recent developments with up-to-date ground investigation techniques. This may create a positive context and the client may decide to order some CPTs in order to optimize the earthquake-resistant design of the LPG storage tanks.

This simplified example presents some of the complexity and interaction of the elements of risk context and risk content between two communicating parties. In practice, even for experienced professionals, it proves to be relatively difficult to separate the rational and emotional aspects of communication. Therefore, Covello et al. (1989) emphasize targeting any risk communication specifically to the receiving person or persons. They stress the importance of paying particular attention to the context of the risk communication. An emotionally loaded risk message to a client, by an upset and stressed design team, is likely to miss its foreseen objectives. Serious attention to not just the content, but also the context, will result in a far better understanding of each other, particularly when dealing with risk issues.

Singularity and risk dominance

Many people experience some difficulty with the application of the concept of *singularity*: the acceptance that two opposite points of view about one factual situation are both true (Watzlawick et al., 1967). For instance, a hard ground layer may be favourable with regard to bearing capacity and unfavourable because of high blow counts required to hammer a foundation pile to the designed depth. Singularity reflects the complexity of our daily working lives by the often conflicting interests resulting from the same factual data.

If there are two opposite truths about one and the same fact, which truth should we select and consider as *the* truth? If we realize that all of us, to some degree, are susceptible to our social environment, this question becomes particularly significant. For instance, we seem to give more attention to the demands of certain people, above other people, as Box 5.2 demonstrates.

Box 5.2   About electroshocks and singularity

I recall an extreme example of the combination of singularity and risk dominance from my MBA lectures. In the 1960s, the psychologist Milgram provided
some literally shocking experiments. The outcome was highly surprising and
the experiments have been repeated a number of times since then, always with
more or less equivalent results.

   For the sake of a so-called medical experiment, a medical doctor requested
a test subject to give electric shocks to another person. These electric shocks
create the singularity for this test subject. They are *favourable* for the experiment
of the medical doctor, but turn out to be very *unfavourable* for the person under
attack during the experiment. The test subjects started to give low voltage
shocks. By the end of the experiment, however, electric shocks of over 300 Volts
were given to the victims, as requested by the doctor in his white coat, for the
sake of science. Typically, most test subjects followed the instructions of the
medical professional, while successfully neglecting the crying of the victims.
Fortunately, the persons under attack by the shocks were fake, otherwise their
deaths could have been expected. The experiments demonstrate how normal
people, like you and me, may be susceptible to socially constructed situations,
with an imbalance of authority and power.

   If we recall the hard ground layer with respect to singularity and risk dominance, how do we deal with it? In its appraisal does the bearing capacity dominate
or the pile driving? Which is the dominating risk? Is it in certain conditions
dependent upon *who* is responsible for which risk? Are people factors, such as
age, experience, and the level in the organizational hierarchy guiding our behaviour? The example in Box 5.3 indicates the possible effects of singularity and its
associated risk dominance within a project team.

   Hedges (1985) emphasizes the importance of group discussions in order to guarantee a shared understanding of the team members' risk perceptions. A facilitator
may lead the discussions, with due attention to the content and the context of
the communication. This approach may help reveal the singularity of particular
ground conditions, as well as the socially-constructed risk dominance within the
team. Moreover, it will reduce the chance of another adverse effect within teams:
groupthink.

Risk of groupthink

Experiments, as well as history, reveal too many examples of the dramatic effects
that so-called *groupthink* tends to cause to people, organizations and even societies.

Box 5.3   A hard ground layer, singularity and risk dominance

Let us consider a geotechnical engineer and a piling contractor. Both are members of our project team under a design and construct contract. The geotechnical engineer and the piling contractor have different opinions about one and the same hard ground layer, which acts as their topic of singularity. The hard layer may be effective from a pile bearing point of view, but adverse from a pile installation point of view.

The geotechnical engineer is responsible for a safe and economic pile *design*. The piling contractor is responsible for a safe and economic pile *installation*. The geotechnical engineer will evaluate the strength of the ground conditions with regard to bearing capacity, the stronger the ground, the better for the bearing capacity. The piling contractor will interpret the ground with regard to pile installation. The weaker the ground, the faster his pile driving. Therefore, they have opposite points of view about the same ground conditions. Within the team, they need to balance safe pile bearing capacity with a smooth pile installation, with a shared understanding of the risks of insufficient bearing capacity and broken piles.

Let us assume that the geotechnical engineer is rather old and experienced, while the piling contractor graduated just two years ago. There is a fair chance that the younger and less experienced piling contractor feels the need to re-value his or her risk perception in favour of the perception of the experienced and older geotechnical engineer. It will be difficult for the young piling contractor to neglect the natural preponderance of age and experience of the geotechnical engineer. Obviously, this socially constructed attitude and behaviour is independent of the fact that the piling contractor's risk perception may very well turn out to be the closest to reality during the project, despite younger age and less experience.

Groupthink rules when team members are unwilling or unable to disagree with one another. In such teams decisions are solely based on keeping the team consensus and harmony (Daft, 1998). In these situations each team member conforms without doubt to the group culture. Independent thinking and asking unconventional questions are not supported and even neglected in these teams, which consequently will create significant tunnel vision. Groupthink supporting conditions are a strong team cohesion, isolation of the team, strong and directive leadership, as well as stressful conditions, without hope of finding better solutions than the leader suggests. Teams with strong and closed cultures are particularly susceptible to groupthink. We may think of sports teams and the army's elite troops. However, even tourists on a holiday-trip may become a victim of groupthink according to Box 5.4.

Box 5.4   Fatal groupthink by Dutch tourists in the Spanish mountains

The Dutch like to travel. Each year many of them visit mountainous regions to experience the pleasure of hiking that is impossible in the flat Dutch landscape. On Wednesday 5 May 2004, a group of 11 Dutch tourists and their guide departed to walk to the top of the Mulhacén, at about 3500 m, the highest mountain in the Spanish Sierra Nevada. They were typically not dressed and prepared for the adverse weather conditions they encountered after one and a half hours of walking: storm and snow.

They planned to reach a hut at 2500 m. Another hour later, the weather was still bad and the group had to struggle in knee-deep snow. Two tourists tried to convince the guide and the other group members to return, because they considered it as irresponsible to continue. According to the guide, it was well possible to continue and to reach the hut to stay overnight. The other group members conformed to the guide.

Because the two persons, one of them an athletics coach and the other a fanatical rower, were not able to convince the group to return, they broke with the group and decided to return to the village in the valley on their own.

The guide with the remaining group continued their hike. During that night a rescue team found them on a slope below the Mulhacén. Earlier, they left two group members behind, who died in the snow. A third member died during that night in the hut, where the rescue team took them for shelter. A fourth member was very seriously affected by hypothermia and went into a coma. Conforming to groupthink or not, in this sad situation it proved to make the difference between life and death.

Tough personalities are needed in order to break through the shared and often fanatically defended opinions of teams that are infected by groupthink. Which is where the self-motivated *individual* plays a key role. Variation with different people with different backgrounds, knowledge and experience, is highly recommended to avoid adverse team effects such as groupthink. Also with regard to risk management, groupthink in teams should be avoided by almost any means.

How teams can contribute

Apart from their inherent complexity, their cultural aspects and their possible adverse effects, created by singularity, risk dominance and groupthink, we have to realize that we need them – teams. They can significantly contribute to effective (ground) risk management, which should result in the highly desired cost reductions and increased profits, within well-defined safety and quality standards. Three types of team are explored in some detail: teams of ground-related experts, multidisciplinary teams and teams as change agents. For effective risk management we need all of them.

Expert teams

Still, too often, the outcome of a geohydrological or geotechnical analysis is presented as *the truth*. At first this may seem logical, because it is the result of engineering activities, supported by figures and calculations. But if we recall the inherent uncertainties of ground conditions, then we may agree with a statement of Barends (2005). He advocates to look for *safe uncertainty* rather than for *unsafe certainty*: 'We need to be confident in our prediction and at the same time suspicious about their large margins. For non-professionals this paradox may render the advice of a geotechnical expert seemingly unreliable'. Table 5.1 provides three examples, derived from the literature, which may clarify what is exactly meant by these margins in geotechnical engineering.

One example by Clayton (2001) compared the observed pile performance with the predicted performance, as provided by 16 different designers. While the observed pile bearing capacity was some 2850 kN, the calculated bearing capacities varied between less than 1000 kN and more than 5000 kN.

Kort (2002) measured the horizontal deformations of a sheet pile wall, while experienced geotechnical engineers provided 23 deformation predictions. Each engineer used exactly the same factual data. The engineers had to interpret the data, to establish the geotechnical design parameters and to select, in their views, the most appropriate calculation models. Their carefully selected design parameters varied with a factor of three through to five. The actually measured horizontal deformation was some 100 mm, while the calculated deformations varied between 50 mm and 500 mm. This range demonstrates a difference of a factor of ten between the maximum and the minimum calculated deformation.

Finally, Koelewijn (2002) demonstrated the effects of using different types of characteristic shear strength values in slope stability calculations. Resulting differences of the stability factor between 0.36 and 1.65 were established, which

Table 5.1: Margins within geotechnical engineering

Geotechnical analysis	Calculated		Measured
	minimum	maximum	
Pile bearing capacity (Clayton, 2001)	1000 kN	5400 kN	2850 kN
Horizontal sheet pile deformation (Kort, 2002)	50 mm	500 mm	100 mm
Slope stability safety factor (Koelewijn, 2002)	0.36	1.65	–

indicates an unquestionable difference between failure and non-failure. Within the same study, five different geotechnical engineers calculated stability factors with margins between 10 and 40 per cent. A hopeful detail forms the fact of decreasing margins with an increase in the number of factual ground investigation data.

The many *choices* which the geotechnical engineer has to make for his or her calculations appear to be the major driving force behind the margins. The *interpretation* of factual geotechnical information and the selection of the geotechnical design parameters are particularly coloured by the inherently subjective *engineering judgement* of the involved engineers. As indicated by Flanagan and Norman (1993), this judgement may be rooted in experience and intuition or gut-feeling. Together, all these engineering choices result in a significant *interpretation difference* between experts. Here we revisit the *interpretation risk*. The existing geotechnical standards and guidelines, remarkably, still allow these subjective engineering approaches in the prevailing practice.

The examples of Table 5.1 highlight the effects of differences in ground perceptions and their impact on the results of geotechnical design. Similar outcomes can be expected for environmental and other types of ground-related aspects, as the underlying principles of ground uncertainty remain the same. In my view, these examples demonstrate the urgent need for a thorough *expert team approach* for many of our ground-related engineering and construction activities. The doubtless dedicated and good-willing calculation of one single geotechnical expert, or even one geotechnical firm, involves the risk of too much bias. These expert teams are well beyond the for many years and widely applied concept of *second opinions*. These opinions do reveal differences in approaches and results, but still leave their underlying causes untouched. The expert team needs to explore these causes to be able to judge their bias and scatter and to arrive at a well-balanced outcome, which explicitly reveals the geotechnical margin to be expected. These teams will probably be able to turn their differences in knowledge and experience into cost-effective and acceptable risk management solutions. In particular, for the more serious ground-related risks, it is highly recommended to establish the team of ground-related experts preferably from different firms or institutes and maybe even from different nations. Only together we will be able to guarantee the safe uncertainty, as has been recommended by Barends (2005).

Multidisciplinary teams

The people working in the construction industry and its ground-related disciplines can be characterized as *professionals*. They are experts in their fields and their firms can typically be classified as *professional organizations* (Mintzberg, 1998). The in-depth knowledge and experience of these professionals is obviously viable, however, their thorough expertise may also limit their view and hinder them from seeing the whole picture of the entire project. Box 5.5 presents one concise example concerning the need to oversee the whole project.

Box 5.5   The need for a multidisciplinary project approach

Let us compare the risk perceptions of two types of engineers in a road project: a geotechnical engineer and a road engineer. The first thinks and designs in dimensions of decimetres or, at best, in centimetres. Settlement outcomes expressed in millimetres do not make sense in the discipline of geotechnics. However, this millimetre scale does make sense for the road engineer and his pavement design. As a consequence, a gap of a factor of 10 to 100 may arise in the design accuracy, and its related risks, between both professionals.

Regarding Box 5.5, is there a rationale for a pavement design with an accuracy expressed in millimetres, while settlements due to underlying soft ground layers can only be expressed within an accuracy of decimetres? At least, both engineers should exchange their opinions about this matter, in order to avoid suboptimum settlement risk remediation measures.

The required *awareness* of this sort of difference in risk perception by professionals can be created in *multidisciplinary* or *cross-functional* project teams. These teams include members from a variety of disciplines, probably from different departments and perhaps even from different organizations. According to Oakland (1993), the main benefit of multidisciplinary teams is reducing so-called *silo management*, which is mono-dimensional management with a reasonable chance that the overall project objectives are suboptimized.

Ground-related teams of experts, as previously introduced, will basically manage their projects from their ground-related point of view. No doubt their ground-related issues are of paramount importance, but other issues are important as well, in order to meet the overall project targets. We may think of the timely arrangement of all sorts of permits or of effective external communication by the public relations officer. Not all ground-experts are ultimately talented and skilled to perform this type of activity, which are often beyond their personal interests as well.

Well-balanced multidisciplinary project teams, with members with complementary skills, can therefore add high value to the project. Every team member can contribute according to his or her *strengths*, and *weaknesses* are compensated by other team members. For instance, an in-depth oriented ground expert may leave the external communication to a team member with the appropriate skills and motivation for it. This approach may mobilize the maximum spirit and power of the team. In my own family team it is my wife who compensates a number of my weaknesses. In this well-working miniature team, I was able to write this book, without becoming entirely isolated from my social environment. As shown in Figure 5.5, any project organization should use both expert teams and multi-disciplinary project teams, which implies some team members join more than one team in order to provide a liaison function between the several teams.

The combined use of different types of team may additionally facilitate the development of learning capabilities. Their cross-functional character encour-ages teams to develop so-called *higher order cross-functional capabilities*. These capabilities cannot be reached by sole individuals,

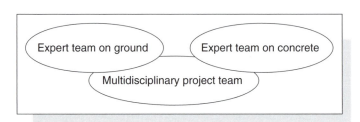

Figure 5.5: Linking expert teams with the multidisciplinary pro-ject team.

which again highlights the importance of teamwork. These capabilities should be chased by our organizations to create sustainable competitive advantage (Grant, 1998). The latter can be defined as creating a persistent higher rate of profit than the competitors in the same market segment. Indeed, it is a highly desirable objective for many engineering and construction companies in today's increasingly competitive construction markets. *Double loop learning*, or gaining new core competencies, by adopting new developments, is needed to arrive at these higher order cross-functional team capabilities. It goes a step further than single loop learning, which is, in fact, only strengthening existing core competencies (Hamel and Prahalad, 1994).

The integration and application of risk management principles in teams typic-ally requires double loop learning. Risk management competencies should be *added* to existing and new construction teams, rather than *replacing* existing know-ledge and experience. While introducing risk management concepts to technical experts, I experience occasionally some unspoken fear that their expert knowledge becomes obsolete. The contrary will occur, the explicitness of risk management will demand, more than ever, sound and state-of-the-art technical expertise to reduce risk to acceptable degrees. Therefore, in my view any engineering and construction team, whether expert or multidisciplinary, should support a mixture of single and double loop learning, because both in-depth and multidisciplinary knowledge are essential for optimum results.

Teams as change agents

Just the enthusiasm of a risk-driven project team is usually insufficient for a successful project result. Such teams need adequate support from their parent organizations and management, particularly by demonstrating trust and openness to the teams. Employees, at all hierarchical levels, should feel free and supported to raise any type of question and to think out of the box. In addition, for truly risk driven (project) organizations, an almost continuous dialogue with clients and other stakeholders is required. This will result in a thorough understanding of the client's and other stakeholder's needs and risk tolerances, as well as of their changes, which supports effective risk management. Therefore, the implementation and application of risk management in our *projects* require a favourable culture in our parent *organizations* as well.

Fortunately, teams prove to be very effective organizational change agents. They may markedly facilitate the process of organizational transformation, as described by a case in the remaining part of this chapter.

GeoDelft, the Dutch National Institute for GeoEngineering, needed to adapt its organizational culture, to respond effectively to the transformations in the Dutch construction industry. GeoDelft decided to become a partner in risk management for its clients. The formation of teams demonstrated to work as a catalyst in this change process. The matrix in Figure 5.6 presents the relationship between four types of deliverables and the required organizational culture and market knowledge (GeoDelft, 2000).

GeoDelft aimed for a strategic shift in its activities, from mainly delivering capacity and products towards particularly delivering services and value. A primarily risk-driven approach of its ground-related research and consultancy activities is one of the services of GeoDelft that adds value to clients. The transformation was realized during the period 2001–2004 and demanded a considerable change

Deliverables	Organisational Culture	Market Knowledge
Value	No hierarchy Network structure External orientation	Strategic knowledge
Services	Creativity Process oriented Less hierarchy	Account knowledge
Products	Formal Procedures and standards Internal orientation	Market knowledge
Capacity	Selling hours No aligned culture	Technical knowledge

Figure 5.6: Deliverables, organizational culture and market knowledge (© with permission of GeoDelft).

in organizational culture towards one with a minimum of hierarchy, a network structure and an external or customer orientation. Existing technical and market knowledge did not become obsolete, but had to be complemented with strategic knowledge of the market. Only with this additional knowledge would GeoDelft be able to provide services and add value as a partner in risk management.

While it is quite difficult to measure a fuzzy characteristic such as the *realized cultural change*, something can be said of the organizational culture change within GeoDelft. For example, the organization became much more customer-oriented. A main driver was the establishment of a number of *cross-functional market teams*. These are responsible for GeoDelft's most important market segments. The members work in different departments and vary largely in experience and education. Following the concept of Piercy (1997), these teams perform *pan-company marketing*, in which marketing is a corporate philosophy and not isolated to the traditional marketing department. In fact, GeoDelft does not have a marketing department anymore. Marketing is executed by the teams, which establish alliances inside and outside the organization, based around customers and supported by information technology. The combination of these factors resulted in a dramatically improved client orientation. The resulting close interaction with clients, on an almost continuous basis, helps to understand better their needs and risk tolerance.

In addition, a lot of employees started to work in teams at project offices of contractors and engineers. Partnerships and exchange programmes were established with contractors, clients and foreign institutes, like Norwegian Geotechnical Institute (NGI), Geotechnical Research Institute (GRI) in Japan and GeoHohai of Hohai University in China. Employee exchange programmes are part of these partnerships, in which GeoDelft employees participated in teams of the partners in an international environment. The new expertise and experiences of these employees were shared and combined to new expertise by *knowledge teams*, in which experts with a shared interest, such as soft soil behaviour or geohydrological modelling, share and build on their knowledge. While knowing *answers* remains important, asking the right *questions* became an essential competence for these teams, to indicate any white spots in knowledge that needs further research.

In conclusion, teams played an essential and even dominant role as change agents in the transformation process of GeoDelft. Together with a number of dedicated individuals, they proved to be the key success factor for arriving at the desired and required risk-driven organizational culture.

Summary

This chapter has revealed and discussed the paradox of the team, which is rather different from just a group of people. Real teams have common objectives, a shared understanding and their members complement each other with skills and experiences. While such teams are not so easily brought to performance, add complexity to the already complicated individual behaviour and require due attention to their culture, they are indispensable in today's construction. Only teams can cope effectively with risk management in our construction projects.

However, to add real value, risk communication within and between teams is of utmost importance. Content and context are two main elements of communication in general and of risk communication in particular. Typical features within teams need to be acknowledged, such as singularity, factual data that result in conflicting opinions, and risk dominance, opinions of more experienced and elder team members that are more easily adopted than those of younger and less experienced members. Furthermore, groupthink, adverse team consensus and harmony that neglects independent and rational thinking, should be avoided by almost any means.

While applying these recommendations, teams have significant contributions to projects. Expert teams, preferably with members from outside the firms and even outside the country, are essential to explore and judge the still wide margins in ground-related engineering and construction. Multidisciplinary teams facilitate our thinking and acting beyond the ground-related issues to serve the overall project goals and to avoid suboptimization. They support double loop learning as well, which is required for the integration and application of (ground) risk management principles in projects and organizations.

Finally, entire teams can also be very effective as change agents. They are able to facilitate the processes of organizational culture change towards more risk management prone organizations that are believed to add value in our highly competitive businesses.

6   Clients, society and risk

Introduction

Most of us work in the construction industry to provide our clients and society with the projects they need, while making a reasonable profit as well. By serving clients and society, we will serve our companies and ourselves too. Eventually, it is all about balancing the interest of the construction industry, to make a reasonable profit, and the interest of clients and society to pay a reasonable price for it. Risk management in general and that of ground conditions in particular can contribute to establish this balance of interests.

After previously considering individuals and some types of team, this chapter explores the relationship of our clients and society with risk. In this chapter we will try to connect the risk perceptions of the project providers to those of clients and society. The latter include the public, people like you and me, who are the end-users of construction projects.

As a first prerequisite for the application of risk management, all parties involved should be willing to make risks explicit and to discuss them in view of their interests. As a second prerequisite, all parties involved need to acknowledge their inherent different risks perspectives, which result from their inherently different interests.

The first part of this chapter explores clients in relation to risk. Differences in perceptions are demonstrated by an example of a contractor's perception of the client and a client's perception of the contractor. Public and private clients are distinguished. The first group represents federal or local governments, while the latter refers to private companies, such as development companies and the oil and gas industry.

The second part of this chapter concerns our society in relation to risk. Topics to be covered are the current post-modern era with its characteristics, as well as changes of risk perceptions over time within society, and the challenge to provide construction projects that really will satisfy the needs of society. As usual, this chapter finishes with a summary.

Clients and risk

An understanding of the clients' risk perception starts with really knowing them and their interests and attitudes. It is of major importance to appraise the *risk tolerance* of the client, the point where the risk becomes intolerable to the client and consequently needs to be reduced or allocated to another party. Clients with a relatively low risk tolerance often choose more conservative designs for construction with lower bounds of risk. In addition, many clients value risk with high probability and low consequences differently from risk with low probability and high consequences, which is likely to result in a more costly design and possibly also in a more expensive construction of the project (Altabba et al., 2004).

Are clients aware of these effects of their own risk attitude? Does our client want to consider risks anyway? Not yet all of them, many clients seem seriously to *fear* risks transparency. This attitude is often caused by lack of expertise and unawareness about risks, specific responsibilities to certain stakeholders, as well as by contractual and financial matters that they are not able to oversee completely.

Do you play the ostrich game?

We may notice a prevailing culture of hunting for perfection in society, which obstructs the adoption of risk management. According to the beliefs of many of us, making failures is not perfect. Risk management displays all sorts of failures and other potential problems. Following one of the main principles of risk management, first we need to become aware of potential problems, only then we are able to take appropriate measures to prevent them from occurring. In other words, first we have to see the *dark side of the moon*.

Many stakeholders in construction, such as clients, contractors, engineers, government representatives, and politicians, hesitate to look at this *potential* dark side of their projects, as experience tends to demonstrate. For instance, they fear that too many explicit risks will prevent their project from starting. They choose to close their eyes and stick their heads in the ground, the *ostrich behaviour*, to avoid seeing and dealing with reality and risk.

It is difficult not to play this *ostrich game* with our clients without losing them. If we do not like to play anymore, there will be many competitors willing to play the game with our clients, which creates a serious dilemma. Box 6.1 presents an example of an often-played ostrich game.

Figure 6.1: An ostrich: not yet playing the game (© photo by Martin van Staveren).

Box 6.1   The ostrich game on settlements

The ostrich game on settlements was illustrated at the symposium *Settlement Requirements – Reasonable and Affordable?* The Dutch Society of Geotechnical Engineers organized it in 2005 in Delft, The Netherlands. The symposium was attended by more than 100 professionals, with one main question: 'Can the contractor realize what the clients asks?' If the client specifies residual settlements of a road of at maximum 0.02 m over a period of 30 years, is there any contractor able to meet that specification at reasonable costs?

Considering the typical Dutch soils with soft clay and peat, as well as high groundwater tables, the answer was immediately agreed by all participants: *no*. As ground professionals, they know these types of specifications are not at all realistic or feasible. However, clients still request these specifications and contractors agree with them, at least at the start of the project, in order to win the contract. The client apparently believes the contractor is able to satisfy his or her pre-set requirements.

As a consequence, both parties will start the project, founded on wrong and unrealistic assumptions. Often, the problems do already arise during construction, when settlements appear to be substantially larger than expected. Who wants to pay for the remediation of these additional settlements? Who has to pay? Who is to blame? Why do we start these unrealistic projects, anyway? What is the role of risk management? Until now, the role of risk management in this type of project is rather minimal, which makes this all happen. It seems not just a typical Dutch disease.

During the same symposium, the attendees recognized the ready availability of improved ground-related technology, better to assess and control soft soil settlements. Innovative theories, as well as geotechnical software packages are able to provide much more accurate settlement predictions. Internet-supported monitoring programmes with easy to perform back-analyses of monitoring data are also available. Nevertheless, this innovative technology is not yet widely applied, simply because there is no contractual drive to apply these technologies. In addition, there also is no clear financial benefit for the contractor to apply these modern technologies, within the prevailing market conditions with severe price competition, in spite of the sharp specifications, as set by the client. It appears to be a rather strange construction world – where is our collective judgement?

Many readers may have their own experiences with ostrich behaviour, for any type of ground-related risk. Why is the ostrich game a problem? It is a problem because is neglects a fundamental project transparency, allows a project to start with unrealistic specifications and often results in a contractual fight between the parties, likely already during construction, with a lot of hassle about differing site conditions, claims, costs of unexpected ground conditions and, not seldom, the end-game at court. How can we break with this type of game without running a serious risk of losing our precious clients? Is risk management an attractive alternative?

A contractor's perception of the client and vice versa

As a next step to move beyond the ostrich game, Boxes 6.2 and 6.3 present examples of the perceptions of a contractor and a client about each other. I derived and summarized both cases from interviews with two leaders in construction in the Dutch magazine *Building Business* (Bijsterveld, 2005; Laverman, 2005). These leaders, representing a government client and contractor, give us some insight about their different opinions, interests, and perceptions. It will also illustrate how risk management can serve as linking pin between these parties. In fact, risk management seems able to bring their interests together, in the direction of a shared understanding.

Box 6.2   A contractor's perception of the client and risk

Improved risk management leads to more client focus, Guus Hoefsloot, CEO
of Heijmans

Guus Hoefsloot is Chief Executive Officer (CEO) of Heijmans, one of Europe's
largest contractors. Mr Hoefsloot connects client focus with risk management.
He argues that it ultimately benefits clients when they realize what sort of
risks contractors may encounter during construction. In case of riskful projects,
selection on lowest price will only result in unsatisfied clients. Therefore, contractors have to become prepared and capable of presenting their risk analyses
to the client.

According to Mr Hoefsloot, there have been a number of major changes in
construction during the last decade. There has been an increase in contractor's
risk, an increase in contractual complexity, an increase in the contractor's
liability, as well as an increase in lump-sum type of contracts. The key for
coping with these developments is risk management. First of all, companies
have internally to professionalize their risk management capabilities. Furthermore, the industry has to share their risk management experiences by exchange
of risk management data and cases, in order to learn from it. Finally, there is
a need for much more transparency in the risk allocation between the client
and the contractor, following the principle that the party which is best able to
manage the risk should become the owner of that risk as well. Of course, this
approach needs suitable procedures and standards of contracting. As contractors, Mr Hoefsloot concludes, we should not try to change the client, we should
change ourselves.

The main message of Box 6.2 reminds us of the famous German duke from
Münshausen, who pulled himself by his own hair out of the swamp. We have
to change ourselves and to become successful dukes of Münshausen, we need to
understand the client thoroughly. In other words, we have to acknowledge the
client's interests from a *client's perception on risk* of which an example is presented
in Box 6.3.

Obviously, Boxes 6.2 and 6.3 provide only two examples from Dutch practice.
However, I recognize similar opinions in many articles and interviews with
change-prone leaders in the construction industry around the globe. I therefore
dare to draw the conclusion that a lot of clients and construction providers are
willing and motivated to share their perception on risk and its management with
each other. In this view, risk management may help to bring clients, contractors
and engineers along side to align and mutually appreciate their main interests, as
much as reasonable, and to create construction projects with a maximum benefit
for all involved. After these examples, let us now focus in more detail on the

---

Box 6.3   A client's perception of the contractor and risk

It takes time to become a good client, Bert Keijts, Director-General of Rijkswaterstaat, the executive branche of the Dutch Ministry of Public Works and Water Management.

Bert Keijts is Director-General of *Rijkswaterstaat*, which is the client of 10 per cent of all projects on infrastructure in The Netherlands. Mr Keijts expects large cost savings from innovative contracts, such as design and construct. He refers to private clients in the oil and gas industry, including Shell, who have already extensive experience with these types of contracting procedures. Another example of an innovative client is the Highway Agency in the UK, which selects the suitable contractor by a fast process, in an early stage of the projects, and in agreement with the rules of the European Union.

By the end of 2007, Rijkswaterstaat will have implemented innovative contracting in its entire organization. This new type of contracting requires new capabilities from both parties, Rijkswaterstaat's employees and the contractors in the market. According to Mr Keijts, not only Rijkswaterstaat, but also the market have still to learn a lot in order to change their traditional attitudes. He wants therefore to start with these innovative contracts in the rather routine and small projects. Within these projects, contractors are expected to be able to oversee their newly gained additional risks. In the view of Mr Keijts, innovative contracts for large and complicated projects will still increase the contractor's risk to an unacceptable level. As a client, Mr Keijts is willing to take some risk as well, when he is able both to recognize and manage them. In his opinion the government needs to be prepared to bear certain risk, for instance to start up Public Private Partnership projects. Finally, Mr Keijts wants more innovative approaches from contractors, as well more interaction with them.

---

particular risk perception of the client. I distinguish the two main types of client: *public clients* and *private clients*. What can we learn from their risk perceptions?

Public and private clients – risk perceptions, costs and income

Public clients typically spend public money, largely generated by taxes, which demands public accountability for reasonable expenditure. Federal, state and local governments are all types of public client. Contrary to private clients, their public equivalents normally do not have direct income from their investment, except toll income from operating bridges, tunnels and expressways. Figure 6.2 schematically shows this situation.

One of the major interests of public clients is expenditure within their budgets. Most clients would try to avoid returning to their local or federal states or ministries to request additional budget, because their construction projects run out of cost control. For that reason, contracting on solely the lowest price criterion, within a pre-set and fixed specifications, is widely considered to

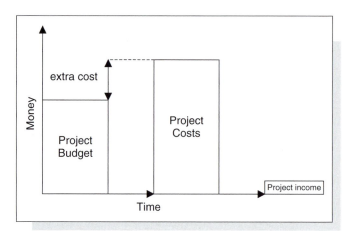

Figure 6.2: Public clients: budget, costs and income.

be the optimum procurement model. The bid with the lowest price, however, rather seldom equals the final costs of the project, as indicated in Figure 6.2.

I have met quite a number of government officials who do not know how to deal explicitly with risks within their conventional way of procurement and contracting. Often, they cannot balance risk and price, as selection on lowest price only rules in their practice. Consequently, numerous risks remain unspoken about. Contractors neglect them as well, in order to arrive at the lowest possible bid price to win the contract. However, the contractor with the lowest bid price needs to absorb a lot of unpriced risks and sometimes even to calculate a negative financial project result. After contract award, particularly when some of the risks effectuate, the contractor will become highly motivated to compensate for at least a portion of his loss by issuing claims. So-called *differing ground conditions* are often (mis)used for this reason. Therefore, despite, or perhaps, because of the lowest price criterion, many public clients still have to request additional budgets. As Figure 6.2 demonstrates, the project income is normally grossly insufficient to compensate for the extra project costs.

Another reason for the lowest price preference of many public clients is the neglect of the *life cycle cost* of the project. Conventionally, public clients operate costs administrations that strictly separate project expenditure and income. Budgets for design and construction are independent of budgets for operation and maintenance, while the lowest total cost of ownership during the life time of a project requires a joint approach of these budgets. For instance, higher design and construction costs may be needed to realize substantially lower operation and maintenance costs, as well as total life cycle costs. A government official needs to explain and defend a higher initial project budget, while a colleague, who is responsible for operation and maintenance, will be cut in his or her budget. That colleague may feel a lowering of professional status from managing lower budgets and probably even fewer personnel involved in maintenance.

Therefore, if we were a public client, who works in the sketched organizational context, why should we make our professional life more difficult, for ourselves and our colleagues? Why should we challenge the status quo? Dedicated individuals and teams, both at the public client's and the contractor's side will be required to break through this conventional thinking and resulting attitudes. Risk management may help these public clients to realize their projects with an acceptable risk profile at reasonable and public accountable costs, with a reduced chance that they have to call for additional budget, sometime during the construction process.

Private clients realize their projects with private money. Examples are investment companies, project developers, as well as the energy and chemical industries. Private clients have to generate income, directly or indirectly, from their investment in construction. Figure 6.3 schematically presents this situation.

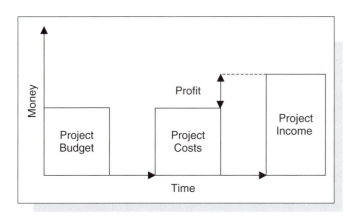

Figure 6.3: Private clients: budget, costs and income.

Compared with public clients, private clients normally operate a much more direct relationship between the costs of the project, the financing of these costs, and the generation of income to make a reasonable profit, as a return on their investment. Private clients are more or less in control of their costs and financing. Their major risk component is normally hidden on the income side. Their income can be highly influenced by difficult to foresee (international) market fluctuations, which is a type of problem most public clients do not have.

Because of their need to generate income, private clients acknowledge the inherent relationship between price and quality. For instance, for many years, numerous companies in the industrial sector have applied additional criteria to the price in their procurement practices. Lowest price is definitely not their sole criterion and for public clients still rather innovative contracts, such as design and construct, are widely adopted and used in these industries. Risk management is not only an integral part of these contracts, it has also been integrated into the daily operations of these industries. The same applies for the life cycle cost approach, which is also more widely accepted with private clients, when compared with their public equivalents. It is one of the ways to maximize income and profits that private parties simply need to satisfy their shareholders and the financial markets.

We can recognize a few major differences between public and private clients, such as the income side and the life cycle costs approach, which may give rise to entirely different risk perceptions between them. However, if their costs are increasing in an unexpected way, for instance by unforeseen and unfavourable ground conditions, both types of client encounter problems in common. Public clients need more budget, which may be politically very difficult. Private clients need more finance and to generate more income, which might be quite difficult when markets are down. Therefore, structured (ground) risk management should be attractive for both types of client.

By now we have gained some insight into the differences and similarities of public and private clients' risk perceptions. I would like to build further on their similarities, by presenting two options for working with our clients.

Confrontation or cooperation

There are two fundamentally different options for working with our clients: *confrontation* (war) or *cooperation* (peace). While most of us are probably quite peaceful people, in relation to our clients we often think that we need to fight. The other way around, for the clients among us, there is a wide spread belief that there is no other way to deal with those contractors than to fight them as well. Box 6.4 provides two examples.

If we want to make a positive difference in the construction industry, by doing it differently, then it is necessary to consider this conventional confrontation model between clients and their providers in some more detail. Therefore, we compare the more traditional model of *conflict* with a more modern approach of *cooperation* or *partnership*.

Conflicts develop because both parties want to gather some benefits, however, from different and conflicting perspectives. Mutually excluding benefits form the very basis of any conflict. It assumes your benefit by definition is contrary to my benefit, and vice versa. In these cases, both *sympathy* and *empathy* are clearly missing. The conflict model hides thoughts about scarcity, the belief that there is not enough for everybody. It is in fact a static and reactive attitude which defends fixed positions that are believed to be under attack. However, if we are able to change our mindset towards a more cooperative model, these attacks may become obsolete.

An alternative for the *model of conflict* is the *model of cooperation*, which concerns *sharing perceptions* by using our possibility to practice empathy for our apparent opponents. It is not needed to feel sympathy as well. That is likely to grow

---

Box 6.4   The model of conflict around ground conditions

Examples of the model of conflict are abundantly available within the construction industry. Each reader will have his or her own stories. One of the main drivers of conflict is the occurrence of so-called *differing site conditions*, when ground conditions prove to be more adverse than expected during construction. Are resulting construction problems, such a severe deformations, caused by differing site conditions or a wrong design? Or is the design correct and did the contractor not sufficiently anticipate the ground conditions, as presented in the ground investigation reports? The answers of the different project parties will vary and are food for conflicts.

The same mechanisms occur in many ground investigation contracts. The client selects the ground investigation contractor with the lowest price. Due to severe competition in a declining market, the ground investigations contractor has to bid below cost price. Once the contract has been signed, the fight starts. Suddenly, locations to be investigated are not accessible with the equipment, as agreed in the contract. Mobilization of appropriate equipment implies, of course, significant additional costs. The fight has started, the price of the bid will in no way represent the final costs to be paid by the client. Their relationship will be damaged and they struggle on.

Experience learned there is a real chance the same client and contractor meet again in their next project. They know each other from before and trust has been reduced to an absolute minimum, on both sides of the table. As an old Dutch saying says, trust comes by feet and goes by horse. Again the lowest price criterion will be applied, market conditions have not been changed and the story becomes predictable. How can we stop this negative spiral?

---

anyway, during the process of working together, as a nice side-effect. A step further in cooperation is *sharing benefits*, in which your benefit becomes my benefit as well.

The model of cooperation is about creating a bigger cake together, so that there will be a bigger slice for all parties involved, contrary to conflict, which implies fighting for the biggest slice. The cooperation model works from the idea of abundance – there is enough for all of us and together we can create even more. It is based on a dynamic and proactive attitude of taking responsibility and catching opportunities by creating economies of scale and knowledge. It supports exploration, side-by-side with our former opponent, in order to create together the best for both parties.

This rather idealistic model of cooperation may appear attractive on paper. However, to put it in practice is a different story. Box 6.5 presents a case about an innovative contract, with the intention to serve both the client and its contractors.

Box 6.5   Attempts to apply the cooperation model

According to Jansen (2001), the client's main objective for applying a design and construct (D&C) type of contract is to arrive at a better combination of price and quality. Priemus (2005) presents additional benefits of reductions in construction time and creative competition on quality. D&C contracts should result in an increase in innovation and dynamics in the construction industry. In The Netherlands, by the end of the 1990s, these advantages were still theoretical and not based on sound practical experience.

The high-speed railway between Amsterdam and Paris is one of the mega-infrastructure projects in recent Dutch history and applied an innovative D&C contract. This resulted in a complete proof of the Law of Murphy, nearly everything that could go wrong went wrong (Priemus, 2005). From a contractual point of view, main problems were:

1   The formation of illegal cartels by the contractors

2   A lack of professional knowledge of operating a D&C contract by the client

3   An overheated Dutch construction market.

According to Priemus (2005), there are several learning points. First of all, the innovative D&C contract type did not fit with the innovative-adverse culture of *both* client and contractors. The culture of forming cartels is an example of being destructive to the objectives of D&C contracts. This demonstrates the need for a pro-innovative culture, with all parties involved, before starting any innovative adventure together. Furthermore, the client was not capable of providing a clear set of functional specifications. The pre-set specifications were of a conventional type and prevented the contractors from providing really innovative solutions.

The entire ground *information risk*, the risk that the factual ground data are not correct or not *complete*, was allocated to the client (Jansen, 2001), which is in conflict with the design responsibility of the contractors. Finally, the complicated interaction of the project with its environment served as an obstacle (Priemus, 2005). Apparently, the soft systems approach was underdeveloped as well.

Risk management was considered of utmost importance, on both the client's and the contractors' sides. I was involved in the implementation of ground-related risk management for a part of the project, when construction had already been started. The contractors' project managers were absolutely convinced about the need and value of risk management. Risk managers were appointed and an extensive risk register was established. However, an awareness about the importance of risk management and, consequently, the individual willingness and dedication to reserve time for its application, was lacking within most of the teams. Risk management was not fully internalized in the entire project organization.

The D&C contract, with all of its consequences, appeared to be still in the early stage of the learning curve for the parties involved. As a consequence, true cooperation appeared to be not that easy. The final costs of the entire project were some 5600 million euros, with a negotiated contract price of 4300 million euros. This implies an almost 40 per cent cost increase for the client, which was the Dutch government. The cost consequences and profit rates for the contractors have not been disclosed. Nevertheless, the aimed cost reduction for the client was clearly not realized by this innovative contract. On the other hand, to highlight the positive side as well, both the client and the contractors showed the guts to start with a new way of contracting for a very complicated and large project.

The attempts to apply the cooperation model, as presented in Box 6.5, turned out to result in a rather innovative version of the confrontation model. However, while not easy, it is possible to apply the cooperation model with our clients. Risk management appears to be an effective tool that facilitates this model, because both risks and the interests of the parties involved become explicitly visible. Later in this book, I will present examples of successful applications of the cooperation models, which have been facilitated by structured ground risk management. Furthermore, Chapter 11 provides guidelines about how to deal with the ground risk in several types of contract, both according to the model of conflict and the model of cooperation. Therefore, we have basically a *choice* between confrontation and cooperation. A combination of both may be possible as well, which is left to the reader to try out.

As set out before, the ultimate goal of (ground) risk management is not restricted to reducing or eliminating risk. It is about targeting the success of our projects. Success is usually connected with creating value. How do we add value to our clients and *their* clients as well? Most clients will be satisfied by either lower cost for the same quality or the cost as budgeted with more product quality or quantity. With such projects results they will serve *their* clients as well, who are often the end-users of their projects and members of the public in society. People, like you and me, who daily use houses, schools, hospitals, roads, railroads, and so on. What about their risk perceptions?

Society and risk

From modern to post-modern risk perceptions

In his classic movie, *Modern Times*, Charlie Chaplin shows his perception of what we can consider as the *modern time*. In one of the scenes Charlie is struggling with monstrous machinery, including some sort of giant clockwork. This struggle seems to be still real.

Our modern times started around 1850, with the industrialization of the western world and a sound belief in linearly forward moving developments. Consequently, today's modern societies focus largely on economic growth, with the annual increase in the Gross National Product (GNP) as performance indicator. However, over the last years, this modern worldview has been subject to erosion. The ongoing globalization and its pressures on a wide range of societies have a large stake in it. Side-effects of fast growing economies, such as increasing urbanization, pollution, and other social and environmental concerns are not taken for granted anymore by an increasing number of people. Critical books about these aspects of modernity, such as *No Logo* by Naomi Klein (2002) and *The Silent Takeover* by Noreena Hertz (2001), became best sellers in several countries. In addition, economies in a number of European countries and Japan have been facing serious economic stagnation for a number of years. Our modern times are changing, as well as the associated perceptions of the public.

Therefore, welcome in today's *post-modern* era. Since the 1980s, this originally rather philosophical term has been established in our language. It took me quite a while to understand its meaning and consequences. The Italian philosopher, Gianni Vattimo, defines *post-modernity* as the existence of *innumerable realities*. The view of one single and objective reality is fragmented into a countless number of realities. These different realities are created by the daily overflow of information provided by the mass media (Vattimo, 1992). Obviously, the Internet facilitates this reality-fragmentation in a dramatic way.

In this post-modern view, *objectivity* and *subjectivity* become increasingly entangled in society, with an unavoidable effect on the way the public perceives risk. Can and do we have to distinguish *real* risks from *virtual* risks? What are their respective characteristics and differences? How do we define them? If a risk is perceived as serious by the public and as insignificant by serious experts, does that risk need to be classified as virtual and irrelevant? If a real risk, in the perception of experts, is seen as only virtual by the public and their politicians, it may not get the remediation it needs to protect that society against it. On the other hand, if a mainly virtual risk is considered as real by society and its politicians, then is management of that risk merely a waste of time and money? Or is it well-spent as it fulfils the demands of the public? This increasing and probably irreversible mixture of fact and fiction is expected to have an enormous impact on the risk perception within society. Fortunately or not, risk perceptions will change over time, also within societies, as Box 6.6 aims to demonstrate.

Box 6.6   Changing risk perceptions over time in society

Around the year 1850, the first train traffic started between the cities of Amsterdam and Haarlem, The Netherlands. At that time there was some panic with the public, in particular with the farmers among them. They perceived a serious risk: cows would not be able to provide milk anymore because of that loud and monstrous steam-driven train crossing the green fields. Currently, there is a dense railway network in The Netherlands and cheese is still one of the important export products.

Two lessons can be drawn from the brief example in Box 6.6. First, risk perceptions within society change. Secondly and very much linked to the first lesson, within society risk perception is often caused by *fear* in its members, which is a normal human response to new and uncommon situations. At the moment new situations arise, there is often a lack of knowledge in the public. Additionally, an open view towards the new situation may be lacking, sometimes within governments as well (see Box 6.7).

Box 6.7   The struggle of a federal government with changing risk perceptions

In the Dutch daily construction newspaper *Cobouw* of 22 March 2005, Professor Jacob Fokkema, Chairman of Board the Delft University of Technology, complained about the government for hiding an open debate with the scientific world about flood risk and construction. In his view, recent views of the Delft University of Technology on the risk of floods to the stability of dikes appeared to be neglected by the government. These new insights were apparently considered as 'unwanted hassle' by the government and improved scientific risk perceptions seem to be difficult to adopt.

Box 6.7 shows that even starting a debate about risk-related topics, with a high degree of public effects, can be difficult in society. Furthermore, the public's trust in experts, who aim to explain how relevant their new insights and developments are, erodes in our post-modern societies, in particular, when other experts express conflicting opinions over the same topic. With today's abundant mass media, these conflicts between experts become quite confusing and frustrating for many people, especially when their opinions affect their daily lives. In this respect I quote Ho et al. (2000):

> ...Finally, our societies are becoming less and less tolerant of failures of engineered structures, including disasters brought about by natural phenomena affecting developed areas. Engineers tend to get blamed for their actions or inactions. There is a pressure for increased accountability and more transparency. Practitioners can hardly hide behind 'expert judgement' or esoteric explanations anymore. Blind public confidence in 'experts' is gradually being replaced by a sense of suspicion: 'It should have been foreseen!'

Balancing all those risk perceptions

As we recognized previously, so many people, so many different risk perceptions. Is it possible to acknowledge and consider all these different risk perceptions in our post-modern society?

Everything surrounding us, our chair, desk, office, house, food and drinks, is the accumulated result of the work of lots of generations of people. Since most of us are not full-time hunters or farmers anymore, our lives depend highly on the activities of other people. Similarly, there is an inevitable interdependency between construction and society. Why is it that so many construction projects are controversial in their society?

Externalities, particularly, can have a large impact on the public's perception of construction projects. This term 'externality' is typically encountered in textbooks about economics and means third-party effects of certain activities in the market (Mulhearn and Vane, 1999). This can involve costs incurred or benefits received by other members of society than the owners and providers of a project. Many construction projects are controversial because they affect individuals and groups of people in an adverse way, while their externalities have not been explained, communicated and reduced in an acceptable way to their stakeholders. Box 6.8 presents how externalities may influence a project.

Independent of the outcome of the additional study of Box 6.8, it shows how the conflict model between the client and representatives of the public has been transformed towards a cooperation type of model, by the power of a hunger strike of one individual.

Obviously, the interests of individuals and groups of people should be balanced with the interests of society as a whole. If we consider each and every individual interest too seriously, most construction projects will never leave the drawing board. On the other hand, when it comes to risk and risk perception in a (post-)modern society, the concerns of its individual members deserve to be taken seriously. Blending all these different risk perceptions to arrive at a successful construction project, as perceived by a maximum number of different stakeholders, is the ultimate challenge for any project manager.

Box 6.8   The impact of externalities on public risk perception in South Korea

One of the numerous examples of the effects of externalities caused by construction projects on the public is provided a 13 km railway tunnel under Mount Cheonseong, some 400 km south of the capital Seoul in South Korea. Environmental pressure groups cited more than 30 protected species and plants at serious risk of disappearance by the tunnel construction. Their perception was at odds with the Ministry of Environment which conducted studies showing out that no endangered species lived on that mountain. Nevertheless, for a 48-year-old Buddhist nun there was ample reason to start a 100-day hunger strike, which stopped the entire construction process of the tunnel. A joint committee of government officials and environmental experts carried out a three-month study to bring the parties together and arrive at a shared risk perception of the environment (NN, 2005).

How do we have to deal with this variety in risk perceptions within the complex social structures that form our societies? One of the ways out is involving all stakeholders, including public representatives, as early as possible in the entire construction process. An open communication about risk with all of these stakeholders should be considered as a pre-condition, in order to make them interested and involved in the project of concern. The stakeholders will acknowledge that their interests are valued and balanced with the interests of other stakeholders. The Australian community of Pittwater demonstrates an enlightened example of structured public involvement in geotechnical risk management, as presented in Box 6.9.

Box 6.9   Public involvement in ground risk management – the Australian way

Like many local communities today, the Council of Pittwater in Australia operates a website with all kinds of information about their community services. Remarkably, it includes the *Interim Geotechnical Risk Management Policy for Pittwater – Council Policy No 144*. This provides a risk management approach for property affected, particularly by landslides, within the Pittwater Local Government Area. Its criteria are based upon the guidelines, as established by the Australian Geomechanics Society (2000). In addition to the policy, a number of forms can be downloaded, such as the Checklist of Requirements for Geotechnical Risk Management Report for Development Application. Those readers interested can visit the website: www.pittwaterlga.com.au.

Another necessity for appraising and acting on society's variety of risk perceptions is flexibility, which allows for project adaptations during the entire construction process. Too often, at least in a highly regulated and densely populated country such as The Netherlands, construction projects are completely cast-in-concrete in their early development stage. Attractive opportunities and alternative solutions typically emerge at unexpected moments, sometime during the realization of the project. These opportunities may satisfy one or more stake-holders and serve society as a whole, but often still remain unused (Versteegen and van Staveren, 2004).

Society, construction and post-modern ground conditions

In the world of ground conditions, the post-modern view is much older than in the social sciences. In fact, since its very early beginning, ground engineering and its related disciplines combine facts and vision, objective ground data and subjective interpretation, to provide optimum ground-related solutions for construction projects. The ultimate and one-dimensional *reality* of ground conditions is non-existent and did never exist either.

Depending on the anticipated use of the ground and the perceptions of the geologists and engineers involved, each ground interpretation includes a largely subjective assessment, as we have discussed thoroughly before.

This inherent subjectivity aligns with our post-modern society and should not be a problem, so long as we are *aware* of this subjectivity. This awareness may, however, be underdeveloped because today's professionals, particularly engineers, are usually trained to apply *technical rationality* (Blockley and Godfrey, 2000). Schon (1983) defined *technical rationality* as a specific type of problem solving, which is based on the technical application of scientific theory. The prevailing education in our societies is largely based upon this principle, because it proved to be successful for the *hard system* type of problems that are independent from the observer. The common scientific laws apply to these problems, such as repeatability under equally controlled conditions. These hard system problems can be effectively solved by using analytical formulas that provide *one* deterministic outcome, according to the outdated modern worldview. However, most problems within construction projects are not independent from the observer, particularly when the observer is the project's end-user.

In many projects purely technical solutions, according to sound technical rationality, do only give suboptimum solutions for the whole project, despite the brilliance of the solution from a technical point of view. This technical oriented solution will not serve the entire project and its public stakeholders, because relevant non-technical factors from psychological, sociological, political or cultural origin are neglected. This situation explains a lot of communication problems between technicians and the remaining members of society.

In order to deal with these situations, Schon developed the so-called *reflective practice* (Blockley and Godfrey, 2000; Schon 1983). It is based on what *works in practice* rather than what is considered as the *ultimate scientific truth*. While *science* is *truth oriented*, *engineering* is primarily *goal related*. Science deals with the absolute truth, while engineering deals with the relative and context dependent truth. In this view, the character of science is modern, while engineering has post-modern features. The reflective practice is based on the dependability of information to its context, rather than on objective truth and precision only, which makes it remarkably suitable to the post-modern worldview on ground conditions and construction. According to the reflective practice approach, any major project decision needs *judgement* from several (public) stakeholder perspectives. Additionally, any expected and required result by the stakeholders should be an integral part of this judgement. Therefore, ideally, for any construction project the reflective approach should be added to the conventional scientific approach, in order to become successfully embedded in society. Awareness of the large subjectivity within the construction practice, together with the recognition of the limitations of the technological approach to problems, will arm us to deal with the post-modern characteristics of construction in general and ground conditions in particular. As a consequence of this inherent subjectivity, success in engineering and construction is much more difficult to measure than success in science, particularly for the non-expert public in society. This may be an explanation why so many construction projects are rather controversial in their societies. Box 6.10 presents an example.

However, it is not all moaning and misery with construction in society. Technically complex construction projects can also be appraised as successful by public stakeholders, as demonstrated by the case in Box 6.11.

These measures of the New York Transportation Authority are comparable to those of the Amsterdam City Authorities in The Netherlands for the North-South Line Underground Project. Together with a lot of public communication, many risk remediation measures have already been taken in the very early stages of this mega-project. One example is operating one of the world's largest monitoring programmes ever, in order to control any deformations of the existing buildings during the preparations and actual construction of the underground metro line.

These latter two examples demonstrate a large public participation already in the early project stages, in fact by the application of some sort of reflective practice. Within these mega-projects the externalities of these projects have been seriously considered and managed. It is our challenge to provide similar approaches to the many smaller construction projects. If we succeed, it will give a major positive boost to the reputation of construction within society to arrive at a reputation it deserves.

Box 6.10   How a technically successful project is perceived as disastrous

The Dutch Betuwe railway project, running from the port of Rotterdam to Germany, can be considered a success story from a technical point of view. Abundant ground-related risk was effectively reduced and construction optimizations were realized by innovative approaches and techniques. One of the project's construction consortia, constructing the so-called Sliedrecht–Gorinchem part, saved some 25 million euros within an innovative partnering agreement with the client. These savings were shared between the partners and part of it flowed back to the government.

However, from a macro-economic perspective the Betuwe railway project is not remotely considered as a success within society. Some may even use the term 'disastrous' from this view point. The operator of the Betuwe route, which starts operation in 2007, expects to operate at break-even level at the earliest in 2012. Until that year, the government should subsidize the railway with 20 million euros *per year* (Schaberg, 2005). Indeed, this amount is approaching the above indicated total savings of the Sliedrecht–Gorinchem part of the railway. Due to this huge macro-economic risk of the government, which is clearly outside the circle of influence of the project consortia, the Betuwe route gained a doubtful reputation from a large part of the Dutch public and politicians. This is a rather sad appendix of a technical success story. Like risk, success is also apparently a matter of perception.

Towards creative construction

The economist, Joseph Schumpeter, introduced the term *creative destruction*. According to Schumpeter, innovation is the central component of competition and the driving force of the evolution of any industry and if these innovations occur at high pace, entire industries can be destructed (Grant, 1998). We will obviously not serve our society by *destruction*, our task is to *construct*.

Let us therefore move from creative destruction towards creative construction. Can we realize a *creative construction* industry? This question was raised by Annemieke Roobeek, professor at Nyenrode Business School and change management consultant, during a workshop in the summer of 2005. Can we give ground engineering and construction some more attractive appeal in our societies? Design clearly demands a lot of creativity, with an important role for the architect. Construction requires much creativity as well, not least because of the inherently uncertain ground conditions. However, until now, the construction industry has not yet been widely recognized as a member of the select group of creative industries. Are we perhaps too much down to earth?

Box 6.11   How a technically complex project is perceived as successful

The East Side Access Project in New York City is possibly the largest transportation project undertaken ever, at least in the city of New York. This some 6300 million US dollars project was initiated by the New York Transportation Authority and will connect Long Island with the East Side of Manhattan in 2012. Geotechnical considerations are a key factor towards the success of this mega project, as well as the communication of its risks with the public of New York (Munfah et al., 2004).

In this project due attention has been paid to all types of environmental issues by extensive coordination with the project's stakeholders and the public. The authorities are fully aware of the fact that a positive public opinion, together with approval of the various affected citizen groups, will significantly contribute to the success of the project.

Therefore, since the earliest project stages, the concerns of these groups have been recognized and tried to be solved by reaching consensus, which can be characterized as operating the cooperation model. Typical concerns are disruption to business activities and nuisance by construction noise, dust, pollutants, and vibrations. An extensive public outreach programme has been effectuated already in the design phase of the project, for coping with these issues. This programme will be continued during construction.

In this respect I cite Barends (2005): 'We must explain to decision makers the uncertainty in geotechnics and raise awareness of the social and economical benefits that derive from risk reduction.' If we are able to blend the application of risk management with innovations, then we may be able to deliver true value to our societies by our construction activities. These innovations can be of various types, such as innovative products, innovative processes and particularly innovative people who make it happen. Only this type of creative construction will be able to deal with all those different risk perceptions.

Summary

The combination of clients, society and risk creates numerous different perceptions of any construction project. These different perceptions are caused by the variety of interests of the many project stakeholders in our societies.

Often, the risk perceptions and underlying interests of clients remain hidden. It appears to be difficult not to play the *ostrich game* with clients without losing them. There are many competitors who do want to play this game of a mutual and conscious denial of risk. To avoid this unfavourable game, we should learn to know about the risk tolerances of our clients.

Differences in (risk) perception have been demonstrated by an example of a contractor's perception of the client and a client's perception of the contractor. Risk management may serves as a vehicle to discuss these differences and it may even facilitate a shared understanding of different interests.

Public and private clients have been distinguished, as both have their different risk perceptions. Contrary to private clients, their public equivalents do not normally have direct income from their project investment. One of their main concerns is to stay within their budgets and many of them use selection on lowest price in their procurement practices, in spite of its adverse effects on risk transparency. Private clients tend to acknowledge the relationship between price and quality, because they need to generate income, directly or indirectly, from their investment in construction. This serves as a basis for life cycle costing and the application of thorough risk management.

Two fundamentally different options of working with clients have been appraised: *confrontation* (war) or *cooperation* (peace). The first is still common practice, the latter is promising, yet difficult to apply in practice as well. Sharing of perceptions and benefits is essential for successful cooperation with a minimum of conflicts.

The members of our societies form the end-users of construction projects. Often the public is the client, or at least stakeholder, of our direct clients. Over the years, our societies transformed from modern to post-modern, in which perceptions in general and risk perceptions in particular became highly subjective, diverse, fragmented, and dynamic in time. Balancing all these different risk perceptions, while including the effects of the construction's unavoidable externalities, is the ultimate challenge for successful and widely accepted construction projects. Creative construction, by combining the benefits of innovation and risk management, may lead to the highly desired construction results which serve both our clients and society as a whole.

PART THREE

The process factor in ground risk management

7 The GeoQ risk management process

Introduction

This chapter provides the bridge from the mainly *conceptual thinking* about the GeoQ *concept* in the previous chapters towards the *practical application* of GeoQ *process* in the forthcoming chapters. It introduces the application of the GeoQ process in our day-to-day construction projects. The key importance of the people factor has been highlighted in the three previous chapters. It is therefore suggested to consider for a while at this moment the following questions, with regard to your own role and those of your team members, client and other stakeholders in your current or next construction project:

1 Are they motivated to make all foreseeable ground-related risks *transparent*?

2 Are they willing to *identify* all foreseeable ground-related risks?

3 Are they willing to *classify* all foreseeable ground-related risks?

4 Are they willing to *allocate* each identified and classified ground-related risk to one or more of the parties involved?

If you expect a loud and clear *yes* on these questions, by each of the persons and parties involved, a giant step forward has been made in the acceptance of ground risk management in the project. If not, some more work has to be done further to convince those still somewhat reluctant persons and parties about the need for risk management.

Anyhow, after this intermezzo we are able to proceed with this chapter about the GeoQ risk management process. First, six GeoQ *steps* are introduced and explained. These will be further explored in depth in the following chapters. Next, six GeoQ *phases* are presented, in which these GeoQ steps have to be made. Each of the chapters 8 through to 13 is dedicated to one of these subsequent project phases. Embedding the GeoQ process within the project organization forms the last part of this chapter, which ends with a summary.

The six GeoQ steps

Six general steps of the GeoQ process provide a conventional risk management procedure that is applicable for any engineering and construction project, big or small, anywhere in the world in any type of ground conditions.

It starts with gathering of *project information*, to provide clarity about the project objectives. Based on this information, risks are successively *identified* and *classified*: what are the foreseeable risks and how serious are these risks? Then *risk remediation* measures are selected and implemented, which may require a serious analysis and even a further breakdown of the already identified and classified risks. After the implementation of these risk remediating measures follows *risk evaluation* of the remaining or residual risks. Finally, all precious risk information of the previous steps is stored in a risk register and *mobilized* to the next phase of the project.

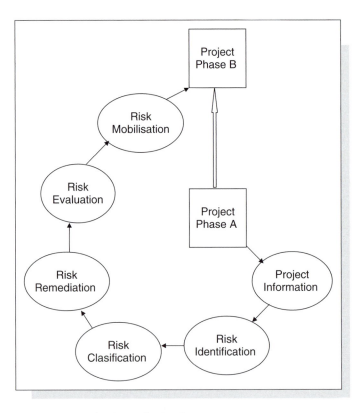

Figure 7.1: The six cyclic GeoQ steps.

Figure 7.1 present these six steps of the GeoQ process from phase A towards phase B in a construction project. Phase A can reflect the design while phase B represents the construction phase.

As stated, for instance, by Macpherson (2001), the initial risk assessment should flow into a continuous risk management programme. Therefore, ideally, these six steps should be repeated in every following project phase. This will provide a cyclic process that facilitates management of the dynamic character of most risks.

Gathering project information

The GeoQ process starts with the *gathering* of all relevant *project information*, in order to provide clarity about the project objectives. We earlier defined risks as obstructions that prevent meeting objectives. Therefore, the objectives to be realized need to become crystal clear in this first step, which serves as basis and context for the following GeoQ steps. We must know the objectives of our project, together with their main characteristics, to be able to manage the project's risks of not realizing them.

As we encountered in Chapter 2, any hard technical system is embedded in its environment, which forms a soft system (Blockley and Godfrey, 2000). Therefore, by following the systems approach, we can distinguish two main types of project information:

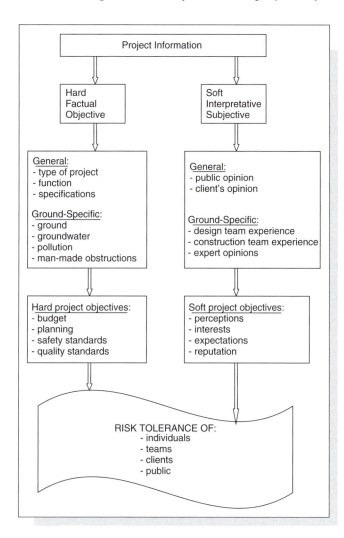

Figure 7.2: From project information to risk tolerance.

1   Hard systems information: factual and objective information

2   Soft systems information: interpretative and subjective information.

Figure 7.2 presents a break-down of both types of information into general and ground-specific information. Ideally, all of this information serves as a basis for the hard and soft project objectives. In other words, the pre-set hard and soft objectives need the support of the available project information, in order to be realistic and achievable. These project objectives, both the hard and the soft ones, will influence the risk tolerance of the project's stakeholders, because the overall risk of any project is not meeting its objectives.

Project specifications can typically be considered as hard and general information. With regard to ground conditions, the maximum allowable differential settlement is a typical example. Hard and ground-specific information can be subdivided in factual ground data, factual groundwater data, factual environmental data and factual data about man-made obstructions in the ground. Maps, boring logs and laboratory test results are sources of this type of project information. Obviously, the availability of this factual information is highly dependent upon the type and the phase of the project. The accumulated hard project data are very useful to appraise the hard and quantitative project objectives, in terms of budget, planning, as well as any pre-set safety and quality standards.

Also the soft project information can be subdivided into general and ground-specific information. Public opinion can be considered a sort of soft and general project data, as presented in Box 7.1.

Box 7.1   Soft project information can be expressed loudly

When a new underground metro-line was planned in the city centre of Amsterdam in the mid-1990s, a lot of frustration about the nuisance of the previous underground metro project surfaced in the local press in Amsterdam. Many inhabitants feared living in a large construction pit for a number of years, as they experienced before. These soft and subjective, but very loudly expressed, signals were taken seriously by the local authorities. This resulted in the design of a mainly *bored* tunnel, instead of a cut-and-cover type, which will minimize the nuisance for a maximum of inhabitants of the city centre during the tunnel construction.

One example of a source of soft and ground-specific information is the experience of a geotechnical project team with the design of a diaphragm wall in a city centre. Another design team may have other experiences. Like the hard project data determine the feasibility of the hard project objectives, the soft project data are of great help to get insight in the soft project objectives of the stakeholders. The latter may include factors like perceptions, interests, expectations and reputation. The legacy of politicians, by the realization of a major construction project, is an example of a soft project objective. While the hard project objectives are measurable, their soft equivalents inherently remain a little diffuse and fuzzy.

The combination of hard and soft objectives provides an indication for the risk tolerance of the project's stakeholders and will markedly influence the next steps of the GeoQ process. Matching these accumulated project objectives with the risk tolerances of its stakeholders is the key to a successful project. Armed with these insights, we are now prepared to take the next step: identifying risks.

Identifying risks

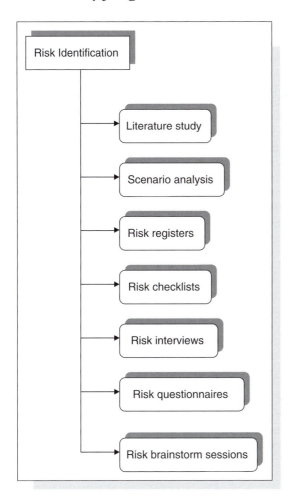

Figure 7.3: Risk identification methods.

Which risks can we foresee for the construction project, in the current and following project stages? Risk *identification* is the second step in the GeoQ process. It is an important step and, according to Baya et al. (1997), the benefits of risk identification are even more important than the next steps of risk classification and risk remediation. This may be caused by the favourable effects of making the unforeseen foreseen, the unspoken spoken and the unwritten written during the process of risk identification. Anything foreseeable that might obstruct the project objectives gets a name and identity by calling it a risk, such as groundwater pollution of a site for domestic housing, rockslides from a steep slope above a new road and excessive settlements of a railroad embankment on soft soil.

However, often the largest damage unfortunately results from the unforeseen risk, rather than from a wrong analysis of the foreseen risk (CUR, 1997). In the mean time, we simply have to accept that the unforeseeable, by definition, always remains unforeseen. In spite of these facts of life, our challenge remains to arrive at a maximum number of fore-

seeable risks and a number of different tools can assist us. If we combine these different tools with as many as economically affordable different risk perceptions, we have done what we reasonably can do and at least the most urgent foreseeable risk will become identified. Figure 7.3 presents a number of widely used risk identification methods or tools that have been derived from a literature research by Viehöfer (2002). The tool of scenario analysis, as suggested for instance by Bles (2003) and Altabba et al. (2004), has been added in Figure 7.3.

The *literature study* as a tool for risk identification probably does not need much more explanation. Numerous textbooks and papers are available worldwide about all types of ground-related issues for engineering and construction activities.

Scenario analysis provides a look into the future, by asking the question *what if* (Thompson, 1997). This type of analysis has been derived from scenario planning that is widely used in the field of strategic management. In the 1960s, the Royal Dutch - Shell group of companies pioneered with multiple scenario development as a basis for their long-term planning (Grant, 1998). It proved successful in the oil and gas industry with its riskful, very long-term and huge investments. In the view of Altabba et al. (2004) scenario analysis requires a detailed description of the project. This forces thinking through the entire project in an early stage by looking towards different types of future. Baya et al. (1997) suggest considering the two extremes, the dream and the doom scenario. It is likely that reality will be encountered somewhere in between. Normally, scenario analysis results in a wealth of identified risks and opportunities. It has proven to be useful to continue scenario analysis beyond the step of risk identification, to support risk classification and to select suitable risk remediation measures.

Risk registers and *risk checklists* are documents with perceived or even occurred risks and their appropriate remediating actions. Clayton (2001) defines the file where risk information is stored as a *risk register*. Apart from a risk description, it usually contains information about the probability of occurrence of the risk and expected risk effects, as well as the ownership of the risk and the risk remediating actions planned or already taken. The British Tunnelling Society (2003) defines the risk register as a *formalized* record of risks and the primary means for recording and monitoring the risk management process. Doubtless, the project-specific risk register must be established right from the start of the GeoQ process. Spreadsheets or more advanced software packages can be used to store all information digitally. The book, *Managing Geotechnical Risk* by Clayton (2001), includes a number of risk registers of leading British construction companies.

Risk checklists are simply lists with a number of foreseeable risks, often related to a particular object. For instance, checklists may be available about the risks associated with hard rock tunnelling, clean-up operations of polluted sites or excavations in soft soil.

Companies or governments may start to create their own risk registers and checklists. It may also become a task of the professional institutions or national institutes. Both risk registers and checklists are helpful during the process of risk identification, as they include the learning effects of previous projects. Risk registers and checklists can also be used in the next steps of risk classification and risk remediation, as many of them may include risk classification and risk control information as well. However, as a warning signal, we have to stay aware of the very project- and site-specific character of risk management. While using risk registers and checklists, independent thinking and judgement remains of utmost importance.

Risk interviews may include individual or small groups of experts. Existing and well-known interview techniques can be used to arrive at a maximum of information. So-called open questions, which start with words like *why*, *what* and *how* can be interchanged with closed questions, which give only answers by yes and no. Many individuals tend to give more information when interviewed on their own than when interviewed in groups. The more different the interviewed people are, the more different their risk perceptions are likely to be, which supports the idea of interviewing both very experienced and less experienced experts with different educational and even cultural backgrounds. Risk interviews can be expanded towards the next GeoQ steps, to obtain information about the risk classification as well.

Risk questionnaires can be used to obtain a lot of identified risks in a structured way from a lot of different experts. Today's electronic mail works very efficiently to reach many experts all over the world with just a few mouse-clicks. As suggested by Keizer et al. (2001), risk questionnaires can be derived from the results of the previously discussed risk interviews. In this approach, risk questionnaires will challenge the experts to add new foreseeable risks of their own experiences. Risk questionnaires can also be applied in the next GeoQ step of risk classification or further.

Groups of people identify risks, based upon a mixture of their individual perceptions and experiences, in *risk brainstorm sessions*. A specific type is the so-called *Potential Problem Analysis* (PPA), with a focus on problem identification and associated risks (Versluis, 1995). These brainstorm groups may only involve ground experts or have a more multidisciplinary character. Which kind of groups are likely to be most effective depends on the phase and the type of the project of concern. Obviously, these brainstorm sessions can build further on the risks already raised in individual interviews and risk questionnaires. These brainstorm sessions can be facilitated by software and so-called Electronic Board Room sessions (EBR). They can be also used to classify risks and to select risk remediation measures.

After considering all these risk identification tools one reasonable question remains: which tools do we need to select for our specific ground engineering and construction project? All that can be said is: *it depends*. A lot of factors, such as the type, size and phase of the project, the anticipated ground conditions, the available budget and time, and the anticipated risk tolerance of the client and other stakeholders determine which combination of risk identification tools provides the best investment. Using a number of the presented risk identification tools will reveal that risk identification is in fact a creative process. It is driven by both hard and soft information from the people involved. Besides experience, which is already a quite subjective factor, also fuzzy aspects such as individual intuition and gut feeling play their (unconscious) role. This may result in a huge number of identified risks of different types and with totally different characteristics. Examples are the earlier introduced pure risks, speculative risks, information risks and interpretation risks.

A common pitfall is to get totally drowned in an overflow of risks at the end of the identification process. If for a particular construction project several hundreds of risks have been identified, which is not uncommon, there is a reasonable chance that the risk management process simply stops, because the people involved cannot oversee and handle it. A simple solution to keep at least some structure during the risk identification process is to store each identified risk in a certain risk group. All ground-related risks can for instance be grouped as geotechnical risks, geohydrological risks, environmental risks and man-made obstruction risks. Another solution to avoid *risk-drowning* is to apply the concept of the risk filter, before starting the risk identification process. By risk filtering, the involved parties agree to exclude certain risk types from the risk identification process. Obviously, these types of risks have to be covered by the competencies of the parties involved, such as sound quality control. For instance, if a piling contractor with a proven track-record and a reliable quality control system will be selected, the risk of failure of a cast-in-place concrete pile may be left out of the risk identification process. The application of risk filtering requires thorough thinking about the construction process in an early stage. In my opinion, this kind of strategic thinking about ground risk management is still in its early development phase. Therefore, in case of any doubt it is advisable to widen the meshes of the risk filter, to minimize the chance of overlooking any relevant risks.

Classifying risk

How serious are our identified risks? All identified risks are *classified* during the next step of the GeoQ process. Risk seriousness can be expressed in the likelihood or probability of occurrence and the likely effect or consequences of the risk. Therefore, often risks are classified in terms of *probability* and *consequences*. Figure 7.4 presents an overview of a number of widely applied risk classification methods.

Viehöfer (2002) proposes three methods for risk classification: *qualitative, semi-qualitative* and *quantitative* risk classification (see Figure 7.4). *Qualitative* risk classification is simply ranking risks in the anticipated order of seriousness. A little more advanced is classifying each risk with terms as high or low, or symbols like + or −. Both the probability and the consequence of each risk can be classified in this fast, easy and straightforward method.

Semi-qualitative or *semi-quantitative* methods for risk classification give both the probability and consequence of a risk a score. These may range from 1 to 3 or from 1 to 5. Multiplying the probability score by the consequence score of that risk gives the risk severity. Risk scores from 1 to 3 will give a minimum score of 1 (very low risk) and a maximum score of 9 (very high risk).

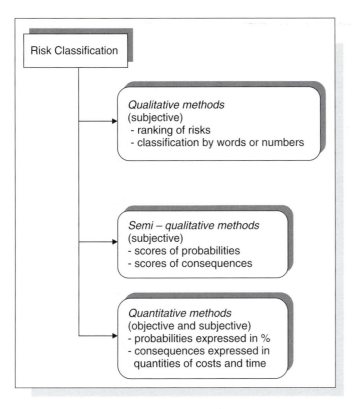

Figure 7.4: Risk classification methods.

Smallman (1999) suggests a more advanced semi-qualitative risk classification method. He subdivides the probability and the consequences of a risk into a number of factors, each with an individual score. The risk probabilities are driven by factors such as the availability of knowledge about the risk and the way the risk occurs, suddenly or gradually. Also a consequence break-down can be made, in terms of cost and time. This method provides additional insight into the nature of the risk, which supports developing suitable remediating measures later in the risk management process.

Finally, *quantitative* risk classification estimates probabilities and consequences in measurable figures. While risk probabilities are expressed in a clear percentage risk, consequences are quantified in terms of costs and time. Engineers and economists tend to like this approach because it allows them to *calculate* risk with these figures. However, there are drawbacks. First, if possible at all, obtaining figures with an acceptable degree of accuracy is mostly expensive and time consuming within ground engineering and construction. Second, if we have them, these highly desired figures are not as accurate as we may assume because they often hide a lot of subjective judgement.

The three subsequently presented methods for risk classification increase in complexity and detail. The required investment in time and cost rises accordingly. Starting any risk classification qualitatively or semi-qualitatively is therefore normally recommended. If required, some specific risks can be classified in greater detail by quantitative assessment.

Altabba et al. (2004) present a slightly different approach in their *subjective* and *objective* risk classification. The *subjective* probability and effect estimate is based on experience and forecasting by experts, which is closely related to the qualitative and semi-qualitative approach. Their *objective* probability and effect estimate is based on statistical records, such as relative frequency records, which aligns with the presented quantitative approach. I have mentioned already the common problem of a lack of sufficient data to allow for statistically sound and objective estimates of probabilities and consequences. However, perhaps we do not need to be too worried about this. As argued by Toft (1993, 1996), Waring (1996) and Waring and Glendon (1998), these *Quantified Risk Assessments* (QRA) may create a false and reduced perception of the real world situation. In reality, risk behaviour is actually much more complex than can be derived from technical knowledge only. QRA might definitely help to judge ground-related risk, as advocated by Ho et al. (2001), however, so long as it is not the sole source for any risk remediation measure. To conclude, for the indicated reasons, I will particularly use the qualitative and semi-qualitative risk classification methods in the GeoQ risk management process.

Remediating risk

After gathering hard and soft project information, identifying risks and classifying them as well, this *risk analysis* needs transformation into action. We have to do something, at least with the serious risks. This next GeoQ step concerns the selection and application of *risk remediation* measures.

However, before jumping into action, we often need some more structure in the freshly identified and classified risks. In addition, a further breakdown and analysis of the risks of serious concern are normally required, before we can select the most appropriate risk remediation measures for them. Figure 7.5 presents these three sub-steps.

Contrary to the previous open-minded and rather subjective processes of risk identification and classification, *risk structuring* and *risk analysis* are particularly rational processes.

Figure 7.5: From structuring via analysis to remediation.

Figure 7.6 presents a number of methods for structuring risks that proved to be suitable in ground engineering and construction practices.

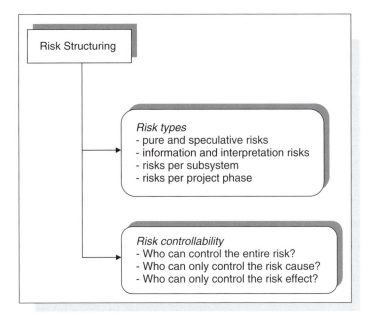

Figure 7.6 content:

Risk Structuring

Risk types
- pure and speculative risks
- information and interpretation risks
- risks per subsystem
- risks per project phase

Risk controllability
- Who can control the entire risk?
- Who can only control the risk cause?
- Who can only control the risk effect?

Figure 7.6: Risk structuring methods.

The easiest way of risk structuring is to group them into certain *risk types*. To get insight into their similarities and differences it is advisable to group risks in more than one way. In Figure 7.6 we can recognize the pure and speculative risk types, as well as the information and interpretation risks. Grouping risks according to the subsystems of a project, such as the risks of embankments and building pits, is another widely used method. Risks are often also related to the process phase in which they may occur. However, care should be taken for the risk on the interfaces of the subsystems and project phases.

Another risk structuring method distinguishes risks by their controllability. Keizer et al. (2001) even add the ability to influence or control a risk as the third main risk characteristic to risk likelihood and risk effect. As indicated by many authors, including Smith (1996), one of the general rules within risk management is that a risk should be the responsibility of the party who is best able to control that particular risk. Useful additional risk structuring is therefore provided by answering the three questions, as presented in Figure 7.6. Based on risk controllability, the optimum *allocation* of the risk may arise. If it is clear which party is, at least partly, able to control a risk, then it is also clear which party should be able to select and implement the most effective risk remediation measures.

Obviously, insight into risk controllability also helps to select the optimum risk remediation measure. However, an almost inevitable step in between risk structuring and risk remediation, particularly for the major risks, concerns the *analysis* of the risk cause and effect relationships. Nor do most risks occur because of one single cause, neither do they occasionally have one single effect and often risks are accelerated by a number of factors. Figure 7.7 shows four proven and widely applied methods for detailed risk analysis.

The *Fault tree analysis* (FTA) is a schematic representation of any *causes* which can result in a pre-defined uncertain event or risk. In other words, FTA aims to identify all possible causes that may contribute to a particular risk. For instance,

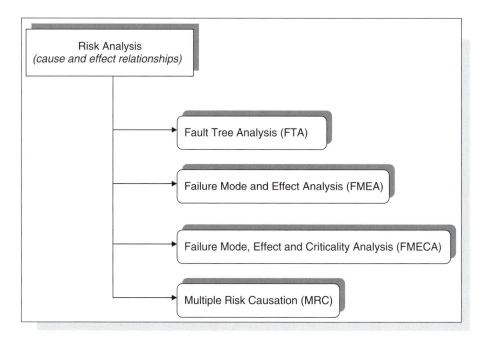

Figure 7.7: Risk analysis methods.

the risk of a rock fall at the foot of a mountainous rock slope can be divided into a set of conditions that ultimately determines the occurrence of the rock fall. Aspects such as the discontinuity spacing and orientation, as well as the water pressures in the rock mass of the slope will all have their contribution to the rock fall. FTA is typically a top-down methodology and it represents the causation of risk.

The *Failure mode and effect analysis* (FMEA) can be considered as the opposite of the FTA. This bottom-up methodology extrapolates the occurrence of a certain event, failure or risk to its possible *effects*. If we return to the rock-fall risk, the FMEA considers all possible *effects* when the rock fall takes place, like obstructing a road at the foot of the slope and causing traffic victims. In addition, a remote village cannot be supplied anymore, and so on.

For reasons of completeness, I also mention the *Failure mode, effect and criticality analysis* (FMECA). This extended version of FMEA assesses and rates both the probability of failure and the severity of its effects, which explain the added term *criticality* (Muhlemann et al., 1992). By building forward on the rock-fall risk example, the FMECA assesses and rates the probability of occurrence, as well as the severity of the effects of the rock fall. Indeed, the FMECA method approaches closely the method of risk classification by scoring probabilities and effects. Due to the structured approach, the FMECA method particularly adds value in those cases demanding a detailed analysis of failures of technical systems. The FTA, FMEA and FMECA methods provide an in-depth insight into both the causes and effects of the risk of concern. In addition, these tools often identify new risks that would remain hidden with the sole application of the previous discussed risk identification tools. Figure 7.8 illustrates a simplified cause and effect analysis for the rock-fall risk example.

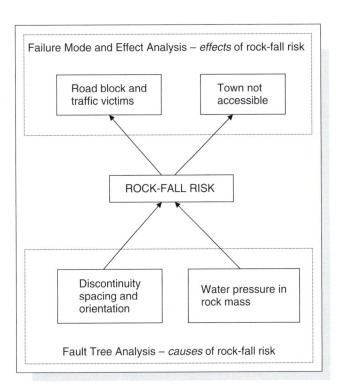

Figure 7.8: Combined cause and effect analysis of rock-fall risk.

The FTA and FMEA methods approach analyses one particular risk by a detailed break down into a number of causes and effects. Acknowledgement of the *causation* of several risks is another significant aspect of risk analysis, because many risks are interrelated. This is what I define as *Multiple Risk Causation* (MRC). A major risk event might occur where apparently unrelated risks line up in a risk chain. Shrivastava et al. (1988) developed a risk model for the avoidance of industrial crises, which gives insight into this interrelationship or causation of risks. Their risk causation framework distinguishes so-called *HOT* failures – Human, Organizational or Technological failures – that are *event initiators*. In addition, they introduce so-called *RIP* factors, which are Regulatory, Infrastructural and Political factors that act as *event accelerators* (Smallman, 1996). The human, organizational and technological failures trigger a certain risk or crisis, while the regulatory, infrastructural and political factors accelerate and increase the risk or crisis. Figure 7.9 applies this approach to the simplified rock-fall risk example and illustrates how apparently very different factors may contribute to the risk of rock-fall.

In this stage, it is of the utmost importance not to get stuck in risk analyses. We should remember that any risk analysis, including preparatory risk structuring, is just a means to arrive at the appropriate risk remediation measures. It is not an objective as such, which I occasionally encounter in practice. In those cases risk management stops when the risk analysis report is ready and issued to the client. Let us avoid this pitfall. After a careful analysis of the structured risks, we are

"HOT" FAILURES (EVENT TRIGGERS)		
Human	*Organisational*	*Technological*
Limited experience of the engineer	No standard design check by a senior engineer	Failing measurement of water pressures in discontiuities
Ultimate Risk Slope instability by a rock-fall		
Regulatory	*Industry structure*	*Political*
No regulation avaible for rock slope design	Severe price competition with selection on lowest price only	None
"RIP" FACORS (EVENT ACCELERATORS)		

Figure 7.9: Multiple Risk Causation for a rock-fall.

well prepared to select the most appropriate risk remediation measures. It is time for action!

Flanagan and Norman (1993) suggest a few widely used risk remediation or risk response strategies. I add another common strategy, which is *risk ignorance*. Figure 7.10 presents these risk remediation measures.

Risk retention is, in fact, just taking or accepting the risk, which may be appropriate for risks with a limited probability of occurrence and limited effects as well. This retention applies also to the so-called residual risks, the remaining risks resulting from risk reduction.

Risk reduction is concerned with either reducing one or more of the *risk causes*, or one or more of the *risk effects*, or a combination of both. In order to select the most suitable risk reduction method, it is useful to get an insight as to whether the causes or the effects of the risk are dominant (de Ridder, 1998). For instance, in the case of a differential settlement risk of a shallow foundation, risk reduction may involve a detailed ground characterization to obtain a reliable settlement prediction (*cause reduction*) in combination with the installation of vertical drains and monitoring during construction (*effect reduction*). The combined application of FTA and FMEA might be helpful in this respect.

Risk transfer or *risk allocation* changes only the *responsibility* or *ownership* of a risk. Without any additional risk remediating action the risk stays the same, so long as all factors of influence on that risk are constant as well – remember the dynamics of risk. The risk controllability exercise may help to decide whether it is interesting to transfer a particular risk to another party. Risk transfer to insurance companies is common for risks with a (very) low probability of occurrence and rather high consequences. The higher the assessed risk probability, the higher the risk premium asked by the insurer will be.

Risk avoidance or *risk elimination* is simply about getting rid of the risk, for instance by stopping the project, or less dramatic, choosing a different design solution. A very high differential settlement risk of part of the high-speed railway from Amsterdam to Paris, has been eliminated by a foundation on thousands of piles. This rather expensive but reliable settlement risk remediation solution has been jointly agreed by the parties involved. If the foregoing options of risk retention, risk reduction and risk transfer are not suitable at reasonable costs, then risk avoidance is normally the remaining option. This is often the case for risks with both a high probability and (very) serious effects.

Finally, *risk ignorance* should not be an option anymore for any reader of this book. In my view, the considerable failure costs in the construction industry result, to a substantial extent, from still widely applied risk ignorance. This ignorance is in no way similar to risk retention because risk ignorance *neglects* foreseeable risk, either unaware or, even worse, in full awareness. It has more to do with the ostrich game and geo-gambling than with responsible engineering and construction. For reasons of completeness, this risk ignorance strategy needs still to be presented.

Obviously, each potential risk response measure needs to be judged on feasibility, costs and likely effects to related risks. Unfortunately, no risk is for free. Each type of risk reduction will incur a certain cost. However, interesting opportunities may arise

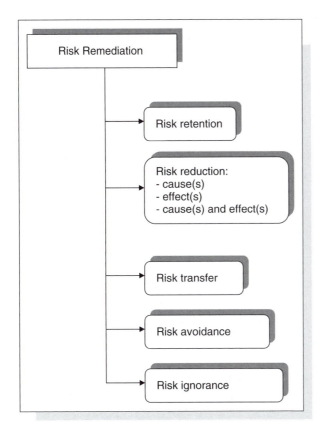

Figure 7.10: Risk remediation strategies.

as well, which may save other costs and perhaps even increase some profits. Selection of the most suitable risk remediation measures is a tedious and complicated task, which should preferably be timely executed by teams rather than individuals, to avoid tunnel visions and over-biased perceptions. Relative outsiders may also be very helpful to put an independent and fresh eye on any proposed risk mitigation measure. Abundant examples of the implementation of risk remediation measures are provided in the forthcoming chapters. It allows us to proceed with the process of evaluation of these measures.

Evaluating risk

After the appropriate risk remediation actions have been taken, the next and fifth GeoQ step can be made, which involves *evaluation* of the resulting risk profile. This step's objective is verifying whether the remaining risk profile is acceptable to the responsible and affected (third) parties, or not. In other words, the *risk response* made needs to be checked and balanced with the *risk tolerance* of the affected parties.

As indicated earlier in Chapter 4 through to Chapter 6, risk tolerance depends highly on risk perception and is dynamic as well. The risk evaluation, therefore, involves two sub-steps. The first is a process check of the risk management procedure applied so far. Have all identified risks indeed been classified? Have all classified risks indeed been remediated in some way or another? Are there any agreed remediation measures not yet executed?

The second part of the risk evaluation is a careful analysis of the (expected) results of the risk remediation measures. This concerns an appraisal of the *residual risks* – the remaining risks after the risk mitigating measures have been taken. Have the risk causes and effects indeed been reduced to the agreed? Regarding the settlement risk example, during this evaluation step the forecasted settlements need to be compared by the extrapolated

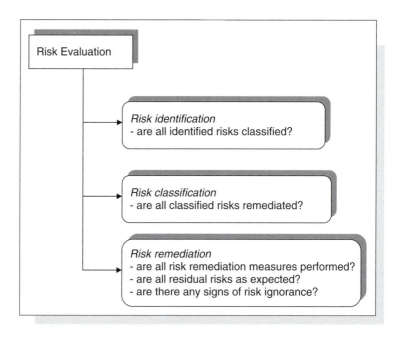

Figure 7.11: Risk evaluation questions.

results of settlement monitoring results. Evidently, the larger the difference between the initial risk and residual risk, in terms of probability and consequences, the more effective the risk remediating measures appear to be. Some main questions that deserve an answer during the process of risk evaluation are summarized in Figure 7.11.

After this evaluation we will probably approach the end of the particular phase in the construction project. We need, therefore, to prepare all risk information ready for use in the next project phase; all relevant risk information needs to be mobilized.

Mobilizing risk information

Finally, in the sixth and final step of the GeoQ process, all gathered and analysed risk management information needs to be properly filed for transfer to the next project phase. The *risk register* has already been introduced and discussed as a risk identification tool. According to the British Tunnelling Society (2003), these registers should be 'live' documents that are continuously reviewed and revised where appropriate. An important moment for such a review is at the end of a particular construction phase. All precious risk information of the preceding GeoQ steps should be prepared, in a suitable format, to be *mobilized* to the next phase of the project. I have deliberately chosen the term 'risk mobilization', to highlight the necessity for *moving* and *arrival* of all relevant risk information in the next project phase. Too often, parties are unwilling to act accordingly, particularly in the contractual phase, which is, in my opinion, a severe threat to the overall transparency of the entire project risk management process. Furthermore, this is a waste of the money previously spent on the risk management process.

We have to realize that by the mobilization of risk information any risk remains untouched regarding its cause and effect, as well as the risk allocation and responsibility. Only the *risk information*, likely by an updated version of the risk register, will be transferred to the next project. Risk *mobilization* is thus no alternative to getting rid of our risks!

The successors of our risk files will be grateful to us if they receive their desired risk information in a decent way. Today, there are numerous approaches to presenting risk, on paper or digitally. Again, I propose the KISS-principle here. We should strive to keep it as *simple* and *short* as possible. In addition, we endeavour to present all risk information as *transparently* as possible.

All these mobilized risk data should preferably serve as a basis in the risk management process of the remaining project phases. Only one GeoQ risk management cycle, as has been presented in this section, will not fully optimize the project's risk exposure. As indicated by the British Tunnelling Society (2003), the *ALARP* – As Low As Reasonable Practicable – risk level will not be reached by risk management during only one project phase. Whether risk management is successful or not depends largely on our ability to keep the cyclic process going throughout the entire project.

The six GeoQ phases

The six GeoQ *steps* reflect a conventional risk management process, applicable for any engineering and construction project, big or small, anywhere in the world in any type of ground conditions. The GeoQ process divides any construction project into six generic and distinct project *phases*. This section presents these six phases and in each of them the GeoQ steps can be applied. That the distinguished project phases reflect a conventional construction process will not bring a large surprise. I am not aiming to be predictable, but I do want to demonstrate the smooth inclusion of the GeoQ process within the day-to-day activities of our engineering and construction projects. Figure 7.12 shows the six GeoQ phases.

Figure 7.12 shows five of the six steps well-aligned. Any construction project will, in some way or another, go through these phases of feasibility, pre-design, design, construction and mainten-ance, during its lifetime. Doubtless, these phases may have other names in various markets and countries. For instance, The British Tunnel-ling Society (2003), together with The Association of Brit-ish Insurers present four phases in *The Joint Code of Prac-tice for Risk Management of Tun-nel Works in the UK*:

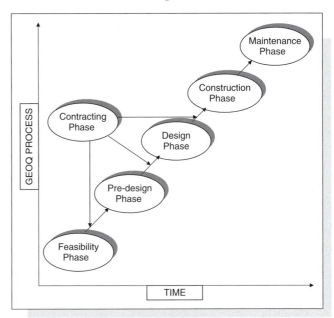

Figure 7.12: The six GeoQ phases.

1   Project development stage – aligns with the GeoQ feasibility phase

2   Construction contract procurement stage – aligns with the GeoQ contracting phase

3   Design stage – aligns with the GeoQ pre-design and design phases

4   Construction stage – aligns with the GeoQ construction phase.

This joint code, which is in use by insurers and re-insurers in the UK, Australia, Canada, France, Hong Kong, Korea, Singapore, Spain and the USA (Knights, 2005), does not explicitly refer to the maintenance phase.

The GeoQ process is entirely independent of a further breakdown in additional project phases for large projects. A combination of certain phases, like joining pre-design and design into one phase, is also possible within the GeoQ process. The six GeoQ *steps*, however, have to be performed in *each* of the distinguished project phases.

One of the six GeoQ project phases, the *contracting phase*, deliberately has a different position in Figure 7.10. As mentioned earlier, in many markets segments in various countries contracts have moved from traditional practices to Design and Construct (D&C) contracting approaches, also referred to as Design and Build (D&B) contracts. Parts of, or even the entire design, are embedded in the scope of work for the contractor in this type of contract. For other projects, maintenance and, even finance and operation, have also been included, which results in contracts like Design, Build, Finance and Maintenance (DBFM) and Design, Build, Finance and Operate (DBFO). Much more of these contractual aspects in connection to ground risk management is explored in Chapter 11. The deviated position of the contracting phase in Figure 7.10 demonstrates the flexibility of the GeoQ process to any type of contract. In other words, the GeoQ process is well applicable to conventional construction projects, in which design is totally provided by the engineer and the contractor just builds, as well as for all sorts of innovative contract.

In each distinguished project phase, the GeoQ process aims to reduce ground risk, as a result of an increase in ground information, compared with the situation without applying the GeoQ process. Figure 7.13 schematically shows the anticipated GeoQ effect.

According to Figure 7.13, the accumulated ground risks would be lower by the application of the GeoQ risk management process, while the total amount of available ground information will be higher than in a similar project without the application of GeoQ. The

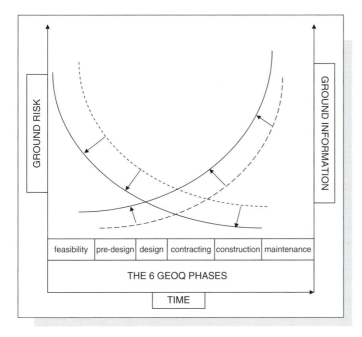

Figure 7.13: The anticipated GeoQ effect.

latter does not automatically involve higher costs for ground investigations and related activities. Within the GeoQ process, any resources spent on the acquisition of ground information are highly focused on the ground-related issues of real (risk) concern, which increases its effects and efficiency and results in a larger amount of information at the same costs. While these benefits are demonstrated by the case studies in Chapter 8 through to Chapter to 13, the six GeoQ phases are briefly introduced below.

The feasibility phase

This book considers the feasibility phase to be the first phase of any construction project. It results in a go–no-go decision about the project's continuation. This decision will be highly dependent upon the financial aspects of the project in relation to its technical feasibility. Also the regional, national or even international political situation and economic developments may influence whether to proceed or not. It is a widely spread misunderstanding, at least in my opinion, that risk-driven ground information data are not relevant factors in this highly uncertain early project stage. Even in this early project beginning ground information is relevant, simply because any construction project literally builds on the ground conditions. It is better to be fully aware of this in an early stage, when it is still possible largely to change or even terminate the project.

Ground risk management in this feasibility phase depends on a number of factors, such as the type, complexity and size, as well as the proposed location(s) of the project. The requirements for ground risk management for a large underground subway in a city centre will be quite different when compared with a small and rather simple project in a remote area. Sometimes, it is already clear that a tunnel has to be constructed, in other situations only a connection between city A and city B has been specified. The main GeoQ purpose in this phase is to provide adequate ground data that contribute to a sound go–no-go decision on the project. Chapter 8 explores in detail the GeoQ process in the feasibility phase.

The pre-design phase

Some sort of pre-design will follow feasibility, at least when the project has proved to be feasible. The pre-design phase normally results in one or a few general and rough design options for the project. The subdivision between pre-design and design is deliberately made in this book. In many projects, the client still wants to stay in control of the first engineering solutions and to decide upon major issues, such as a cut-and-cover or bored tunnel, before bringing the project to the market.

During this phase several engineering solutions are considered and compared. GeoQ is used as a risk-driven facilitator in the process of identifying the optimum engineering solutions, in view of the anticipated ground risks and opportunities.

The optimum solution will be selected at the end of the pre-design phase, based on the specifications of the project, the client's wanted risk profile, and the expected ground conditions.

Then the client will continue a more detailed design or bring the project to the market. In the latter situation the contracting phase will be next. Chapter 9 explores in-depth the GeoQ process in the pre-design phase.

The design phase

Conventionally, design follows pre-design, either initiated by the client or performed by a contractor in a design and construct type of contract. During the design phase the selected solution is worked out in detail, in order to start construction. For instance, after deciding a bored tunnel for a river crossing, this solution will be worked out in detail and the bored tunnel design will be ready for construction by the end of this phase.

In this book detailed design for construction is considered as part of the design phase, while recognizing that a lot of the larger projects usually separate these two phases. As mentioned, this will not influence the GeoQ approach, apart from the fact that the entire GeoQ process should be performed in both phases, including all six steps. During design and detailed design the GeoQ process provides risk-driven engineering support, within the project specifications, the risk tolerances of the parties involved, and the expected ground conditions.

If there is a contracting phase in between design and detailed design, implying the contractor performs the detailed design, then it is of particular importance contractually to allocate all ground-related risks. Any contractor deserves to start activities with a maximum and accessible insight into the existing risk registers of the project. By the end of the day, both the client and the contractor are expected to benefit from this type of transparent risk management. Chapter 10 explains the GeoQ process in the design phase in more detail.

The contracting phase

Contracting follows the design phase for conventional projects. To an increasing extent contracting takes place after the completion of pre-design or directly after the feasibility phase. The client selects and contracts a suitable contractor, who will realize the project. Operation and maintenance is sometimes included, for a number of years. The result of this phase is a hopefully clear contract between the client and the contractor, with transparent and acceptable conditions, for both parties involved. If not, then this phase is likely to be the start of a fight between these parties, often with a lot of disagreement about ground conditions probably in a dominating role. There are seldom real winners after these construction fights resulting from the inevitable cost increase.

Combination with the Geotechnical Baseline Report (GBR) makes the GeoQ process particularly interesting in this contracting phase, because it provides a practical tool for the clear allocation of all types of ground-related risk. Only then may we expect ground risk to be really *managed*. Each risk should have one or more owners, who are contractually responsible for that risk. Any risk can become the sole responsibility of one of the project parties or is shared. This approach proved at least to reduce the claims about all kinds of perceived different ground conditions, as is explored in detail in Chapter 11.

The construction phase

Depending on the type of contract, construction follows either the contracting phase or the design phase. Construction should result in a well-completed project, preferably within budget and planning and according to the pre-set safety and quality standards. Depending on the appraised project results, which includes a large portion of subjective perception as previously has been described, the completed project may have either a positive or an adverse impact on the reputation of both the contractor and the client.

During construction, GeoQ supports a smooth, i.e. efficient and effective, construction process from a risk-driven ground conditions point of view. Ideally, also in this phase all of the six GeoQ steps are performed, particularly because construction is the proof of design quality, including the anticipated ground conditions. Also in construction, the proof of the pudding is in the eating. Many types of monitoring proved to be of great help in minimizing construction risk, as well as to catch unforeseen opportunities from better ground conditions than expected. Examples are ground strength higher than foreseen and pollution concentrations less than anticipated. An innovative contractor may use these opportunities for increasing the project's profitability, by the application of innovative approaches to catch the formerly hidden ground surprises. Chapter 12 focuses on the application of the GeoQ process during construction.

The maintenance phase

Sooner or later, during the lifetime of its operation, the completed construction project needs some sort of maintenance. Sometimes this maintenance phase is part of the contractor's scope of work, for example for 10 or 20 years after completion of the project. The rationale for this type of contracts is a more sustainable design and construction, which reduces the total life cycle costs for the client. In addition, it is in the interest of the contractor to reduce the maintenance costs within the entire scope of work, in order to maximize profits. This may also be of interest to the end-users of the project, such as a road or railway, because they will experience less annoyance by reduced maintenance activities.

It may be helpful also to perform the GeoQ steps in this phase, in order to obtain insight into the behaviour of the construction during its operation with regard to ground behaviour. In addition, the results of the GeoQ process may help to dissolve any disputes in a mutually effective way, in the case that the project's behaviour differs from the agreed quality standards as a result of unexpected and adverse ground behaviour. Again, like during the construction phase, also in the maintenance phase, many types of monitoring contribute to sharpen insights into the object's behaviour and its optimum maintenance programme. Finally, Chapter 13 explores the GeoQ application in the maintenance phase.

Embedding the GeoQ process

Finally, the last section of this chapter allocates some words to the implementation of the GeoQ risk management process within the project organization, based on experiences from colleagues and myself. Generally speaking, if there is not sufficient support in all ranks of the project hierarchy, it will be most unlikely that the full potential of GeoQ ground risk management arises. This means that we are likely to miss opportunities of cutting costs or adding more value to the project. It may pay off to get risk management embedded in the genes of the most important project players. In the words of Findlay Macpherson (2001), finance expert of *International Construction*: 'The cost of employing an independent engineer to provide the risk overview (as many project lenders now insist upon) is real value for money when set against the project cost, and more so if cost and time overruns are prevented.'

However, a common pitfall is to establish a *separate* risk manager in the project organization, who is responsible for the entire risk management of the project. How wrong this proved to work out! It is experienced that many, if not all, people involved in the project point to the poor risk manager when there is anything to do with risk management. They expect the risk manager to solve *their* problems, which typically belong to *their* own responsibilities. I recall the basic risk management rule about who has to manage which risk. Within any project organization, the application of this rule implies the risks and even certain risk causes to be managed by the people who are best able to do so. Normally these are the well-known professionals in every project organization, such as the project director or manager, the design manager, the construction manager, the site superintendent, the quantity surveyor, and so on. The role of any risk manager should be restricted to coordination of the risk management process. Perhaps it is even better to avoid the use of the term 'risk manager' in the project organization, to avoid any confusion.

The overall responsibility for risk management needs to be allocated to the project director or project manager. Integrating risk management in the entire project organization is typically a challenging responsibility for these persons.

They have to make risk management a normal and accepted part of the day-to-day work of all involved in the project, like the daily registration of the working hours is an accepted part of anybody's work.

However, because risk management is still rather new to many (project) organizations in the construction industry, this structured implementation of risk management in the daily activities proves to a rather heavy management responsibility, as indicated in Box 7.2.

Box 7.2   Attempts to embed GeoQ

One of my colleagues was working as a ground risk coordinator at a site office of a design and construct consortium of contractors that constructed a part of the high speed railway from Amsterdam to Paris. I had the opportunity to attend a number of meetings with him and the management of the project organization. The project director previously worked in the offshore oil and gas industry, and he proved to be a real advocate of risk management. So was the management team of the project organization. However, it proved to be very difficult for the risk coordinator to keep the highly required risk management attention alive within the different project teams of designers, planners and builders. There occasionally was a lack of time, or more a lack of priority, with many of these employees, not due to any wrong intentions, but simply because their full awareness about risk management, with all its hidden benefits, proved to be underdeveloped. They simply were not yet able to see their own and their organization's benefits of risk management. Therefore, from their perspective it seemed quite logical not to dedicate that much time and energy to it, in the heat of their daily construction work on site.

Unfortunately, I do not have any quick wins or easy solutions to deal with this type of risk management implementation problem. In fact, we are entering the discipline of *change management*, which moves beyond the scope and focus of this book. As a pioneer and change agent on ground risk management, you are likely to encounter the many obstacles as described in the abundant books about change management. For your necessary inspiration, and to compensate for the unavoidable transpiration, for instance the works of Cameron and Quinn (1998), Jick (1993) and Senge (1990) may be of assistance. Box 7.3 presents a few recommendations, which merged from a mixture of book wisdom and my colleagues' and my own experiences.

The recommendations in Box 7.3 may give some direction during the challenging process of embedding (ground) risk management in the project organization.

Additionally, we should remember risk management not only concerns managing risk, but also supports and facilitates innovations. In this respect the following statement by Manfred Nußbaumer and Konrad Nübel (2005), made in their paper *Portfolio Based Approach to Project Risk Management*, may motivate: 'Transparent organised risk management and well-structured display of the risk situation of a project is always the key element in technical development'.

Box 7.3   Recommendations for the implementation of the GeoQ process

- Do *not* appoint a separate risk manager, use a consultant to train and implement the risk management process, if required, or appoint a risk coordinator, but make risk management explicitly an integral part of the usual management tasks and responsibilities, such as project management, design management, construction management, quality management and safety management

- Start as soon as possible with the risk management process and reserve ample time and resources to teach the key players in the project about its people and process aspects

- Do not accept any concession to any agreements made about the operation of the risk management process, be very tough about this topic, the process needs to be continued to the end of the project

- Read a few books about change management and translate some of their lessons to your own risk management practice

- Stay dedicated and patient, it may even take another generation of construction professionals before risk management is really embedded in the genes of construction professionals and will be developed to its full potential.

Summary

This chapter has introduced the GeoQ process for ground-related risk management, by its six GeoQ *steps* and its six GeoQ *phases*. The first step, gathering *project information*, provides clarity about the project objectives and creates insight into the risk tolerance of the project's stakeholders. Based on this information, risks are successively *identified* and *classified* by steps two and three. Step four involves taking *risk remediation* measures, after risk structuring and a careful analysis of the causes and effects of the major risks of concern. Then follows *risk evaluation* of the remaining or residual risks, by GeoQ step five. Finally, in step six all the precious risk information of the previous steps is stored in a risk register, often supported by databases, and *mobilized* to the next phase of the project. For each of these steps a number of tools and methods have been introduced. We will encounter these in much more detail, including their use in numerous cases in practice, in Chapters 8 through to Chapter 13.

Six general project phases, in which the GeoQ process steps have to be applied, are feasibility, pre-design, design, contracting, construction, and maintenance. The main objectives of these phases, as well as the role of the GeoQ process, have been concisely presented. Both the GeoQ steps and the GeoQ phases include conventional, proven and flexible risk management practices, which allows their application to all types of projects and ground conditions, anywhere in the world.

However, in spite of its flexibility and foreseen benefits, embedding GeoQ in the project organization proves to be rather complicated, similar to many other change initiatives in (project) organizations. If there is not sufficient support in all ranks of the project hierarchy, it unfortunately will be most unlikely to catch the full GeoQ potential, which means missing interesting opportunities of cutting costs, adding more value, or raising increased profits from the project. Some proven recommendations may guide the reader along this large pitfall.

8 GeoQ in the feasibility phase

Introduction

Feasibility normally is the very first phase of any construction project and Figure 8.1 shows the feasibility phase at the very beginning of the GeoQ process.

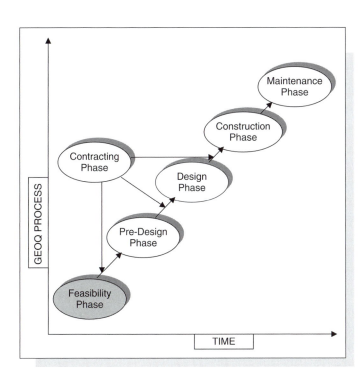

In the feasibility phase, the GeoQ process particularly concentrates on the first two GeoQ steps of *gathering information* and *risk identification*. Ground conditions can be dominant and may even act as the so-called *fatal flaw* to the project. We may think of a large dam planned on a geological fault zone in an earthquake-sensitive area. In such a situation the project needs to be entirely relocated or even cancelled, because of the unacceptable ground conditions.

This chapter starts by introducing and discussing three approaches that support ground risk management in this earliest project phase. These approaches are related to the two GeoQ steps of *gathering information* and *risk identification*. The first approach

Figure 8.1: The feasibility phase within the six GeoQ phases.

applies to the first GeoQ step and classifies the project site with regard to the four main ground-related risk types. The second approach presents how to do more with a minimum of ground information from a ground risk management perspective and the purpose and advantages of scenario analysis forms the third approach.

After discussing these three methods two case studies are presented, in which the presented ground risk management methods have been successfully applied. As usual, the chapter finishes with a summary.

Ground risk management during feasibility

Site classification

Ground risk management in the feasibility phase depends on many factors. Probably the most important are:

- The *type* of project

- The project's *complexity*

- The *size* of the project

- The project's *functional specifications*

- The proposed *location(s)* of the project.

Information about the project's type, complexity, size and functional specifications will be readily available in most cases. This may be quite different for the characteristics of the proposed location(s) of the project. For some projects just a wide region has been selected somewhere in which the project should be located, which typically occurs for infrastructural projects in remote areas. The location for other projects is more or less, or even totally, fixed, such as the extension of a new metroline in a city centre or the development of a new urban area at the boundary of an existing city.

Any decision upon the project's feasibility demands a certain minimum amount of information, including the proposed project site(s). Three main types of project site are distinguished: *greenfield, brownfield* and *greyfield* site. For an effective gathering of project information, first considering the main site characteristics will be highly supportive. Figure 8.2 presents the key characteristics of these three types of site.

Please be aware of the *dominating* character of the ground aspects in Figure 8.2. Obviously, the four major types of ground risk – geotechnical risk, geohydrological risk, environmental risk and risk caused by man-made structures – will surface in most projects to a certain degree. However, in a real *greenfield* site ground and

Type of Site with Main Characteristics	Dominating Ground Aspects	Examples
Greenfield - outside cities and in remote areas - project location not fixed - no existing structures - pollution unlikely	*Greenfield* - geology - ground and groundwater conditions	*Greenfield* - Three Gorges Dam, China - Palm Island, Dubai
Brownfield - often near or in cities - project location is fixed - existing industry remnants - pollution likely	*Brownfield* - pollution of ground and groundwater - existance of man-made structures in the ground	*Brownfield* - Sites for Olympics 2012, London - Redevelopment Projects, Tokyo
Greyfield - in city centres - project location is fixed - existing structures in use - pollution possible	*Greyfield* - interaction with existing structures - existance of man-made structures in the ground	*Greyfield* - Ground Zero, New York - Pudong Business District, Shanghai

Figure 8.2: Three types of project site.

groundwater pollution or man-made structures like buried pipelines are unlikely to be encountered. The geology will dominate in these types of projects, with the associated geotechnical and geohydrological risks. It may be possible to relocate the project to an area with more favourable geological conditions, for instance to avoid the active fault zone for a dam.

For *brownfield* sites the pollution of ground and groundwater is often dominant, in particular, when former industrial sites are redeveloped for new functions, such as domestic housing. These situations occur in Japan for instance, where industries move to China and their sites become available for other functions.

The existing built environment dominates the *greyfield* sites, with normally numerous structures and services in full operation. Greyfields are in fact not real

fields anymore and huge underground metro projects or massive high-rise buildings in cities are typical examples of these sites. In addition, also quite smaller
projects, like an office building or a subsurface parking lot with a construction pit
closely located to a historical building, are typically realized on greyfield sites.

By distinguishing between greenfield, brownfield and greyfield project sites, the
ground information to be gathered for early ground risk management in the feasibility phase can be focused to do more within minimally available time and budget.

How to do more with a minimum of ground data

Most of the ground data are fuzzy and random. In addition, ground data are
by definition incomplete. This incompleteness is at a maximum level in the early
feasibility phase of any project, when site investigations normally have not yet
been performed. We have to deal with the scarce ground data that we can retrieve
in this stage, at reasonable costs, and we should make the best of it. Therefore, the
second risk management approach during feasibility concerns how to do more
with a minimum of information from a ground risk management perspective.
This supports the first GeoQ step of *gathering information* and the classification
according to greenfield, brownfield and greyfield sites is followed.

True *greenfield* sites are often located in remote areas. Distance, a lack of existing
infrastructure and harsh climatic conditions, which may vary from desert to polar
circumstances, usually make it difficult, if not impossible, to obtain a reasonable
amount of ground data. However, some data may be available, like geological
maps, hazard maps, aerial photographs, and satellite pictures. Sometimes, an
engineering geologist, who performs some simple field tests and brings back a
few ground samples for index testing, visits the selected region. Depending on
the morphology and topography of the area, the engineering geologist can obtain
precious ground data from such a site visit. Rock outcrops provide invaluable
data on the ground conditions, which is much more difficult to reveal in rather
flat areas that are covered by soil and vegetation. The type of vegetation may
disclose some possible subsoil characteristics. Furthermore, project experiences in
areas with a similar geological setting may be of great help, while acknowledging
the fact that the actual local site conditions can be quite different from the first
estimates that are based on apparently similar or comparable sites. This early
project phase with a greenfield site will be a real challenge for an adventurous
engineering geologist. Figure 8.3 shows a picture of a typical green field site: a
remote area in the Middle East, where for instance a road is planned.

The acknowledgement of the inherent scarcity of ground data in the early phases
of greenfield projects is a first step to a risk-driven project approach. Clients,
particularly, need to become aware of this simple reality and have to accept the
associated risks of taking wrong decisions, because of the limited ground data
available.

Figure 8.3: A typical greenfield site.

Brownfield sites are located in or near cities and are typically industrial areas. Many of these are (partly) abandoned, often for a number of years. Pollution of ground and groundwater is highly likely in these areas. The awareness of pollution, caused by industrial activities, has developed during the last decades of the 20th century. The expected type and concentration of pollution is highly dependent upon the type of industry and its history. Sometimes, industries have been operating for a hundred years or more at these locations. The rather new discipline of geoenvironmental engineering largely contributes to gathering suitable ground data in this early project phase. Archives of local governments may reveal (indirectly) valuable indications about the type and concentration of pollution to be expected. Geotechnical and geohydrological data can be present to some extent, because of foundations and other elements of the former industrial structures. However, these data are probably incomplete and out-of-date, due to major changes on the site during the industrial activities.

As for greenfield sites, for brownfield sites the amount of relevant ground data normally are of limited extent. In particular, a lack of knowledge about the concentration and type of pollution involves major risk for any client, as environmental clean-up programmes vary dramatically in terms of investment and time required. In addition, as raised by Quint (2005), within the existing regulations and guidelines there is a lot of confusion about what exactly is 'contaminated land'. Figure 8.4 shows a photograph of a brownfield site in south-east Asia.

Such a site will be a real challenge for an engineering geologist with geoenvironmental knowledge and experience to get a maximum of relevant ground data on the desk in this early project stage. Again, the acknowledgement of the restricted ground information facts is a major first step towards a risk-driven project approach. The parties involved have either to accept the situation with the associated risk, or to invest further in gaining some more data, to take the required decisions in this early project phase.

Greyfield sites are typically located in a built environment; the project location is more or less or entirely fixed. We have to deal with the local ground conditions, these are a given, whether weak or strong. In this respect the following words of Heinz Brandl (2004) may inspire: 'There are no (insurmountable) weak soils or rocks, there are only weak engineers'. By the way, he adds to minimize this weakness by education, training and gaining experience.

Another reality for greyfield projects are the existing structures, located at, below, above and around the proposed project site. These structures are still in operation or service and may be very sensitive to the slightest construction

Figure 8.4: A typical brownfield site.

work. Some good news; there is probably quite a lot of existing geotechnical, geohydrologial and may be even geoenvironmental data, once applied to the realization of the existing structures. These data may be still up to date in rather new development areas, like the Pudong Business District in Shanghai. In older city centres the available data will be more obsolete, because of age, incompleteness and serious changes after the data have been retrieved. Anyhow, some ground data should be available for greyfield sites and in several urban areas even databases with ground information are available, in which we are able to select ground data on-line via the Internet, just by entering the postal code of the project area.

Figure 8.5: A typical greyfield site.

The main challenge for a greyfield site is to arrive at some reasonable judgement about the feasibility of the anticipated project in view of unfavourable effects on the existing structures. Monitoring results and photographs of existing structures may help. Indeed, this serves a real challenge for a geotechnical engineer with a sound structural engineering background, to retrieve a maximum of ground data in this phase of the project.

As for greenfield and brownfield sites, also for greyfield sites the amount of relevant ground data are normally still too limited. A lack of knowledge about the interaction of new and existing structures can particularly involve major risk for the client and any third parties

involved. Figure 8.5 shows a picture of a typical greyfield site, close to the Petronas Twin Towers in Kuala Lumpur, Malaysia.

Imagine the consequences, regarding safety concerns, disputes and claims, in the case of unacceptable differential settlement of a high-rise office building, because of a deep construction pit in its close vicinity. Such events have a dramatic impact on the project and its stakeholders. Particularly for greyfield projects, the appraisal of this type of risk is an important first step towards a risk-driven project continuation. Clients should be convinced about this situation and be willing to accept the associated risks.

The gathered ground data at either greenfield, brownfield or greyfield sites have to be reported in a clear way, preferably with a sound motivation for the man-made and natural hazards to be faced. Factual data should be clearly separated from interpretative data, to allow distinction between information risks and interpretation risks later in the process. Major uncertainties should be expressed explicitly, like those of geological origin (greenfields), geoenvironmental origin (brownfields) or the interaction between existing and new structures (greyfields). All of these data can be well used as a basis in the next step of the GeoQ process: risk identification. Scenario analysis is one of the tools for this.

Scenario analysis

If we have classified our project site as either greenfield, brownfield or greyfield and have gathered all available ground data, then we are well prepared to take the next step of the GeoQ process: *risk identification*. For two reasons scenario analysis seems very suitable as a risk identification tool in this very first project phase:

1 The still very *open character* of a construction project in this phase. Apart from functional specifications and rather general project characteristics, such as purpose, size and possibly the location, not much more has been cast-in-concrete yet. There are ample possibilities for all kind of ground engineering and construction solutions

2 The *restricted availability of ground data* with respect to geotechnical, geohydrological and environmental conditions, as well as the presence of any man-made and subsurface obstacles.

As an attractive characteristic, scenario analysis combines these two aspects and with some creativity it is very possible to create a number of interesting and realistic project options, based on a limited data set. Courtney et al. (1997) distinguish four basic types of scenario. Box 8.1 presents these scenarios, together with my translation of them towards engineering and construction projects in the feasibility phase.

Box 8.1   Four types of scenario

1   A *clear enough future* is like a straight road from town A to town B, based on complete information, which is impossible in our practice of engineering and construction with its inherent fuzzy, random and incomplete ground information

2   *True ambiguity* is the opposite from the clear enough future. There is no basis for any forecast about the final format of the project. It is not as bad as that in construction, so also this type of scenario is unlikely to be of use to us

3   *A range of futures* is like a continuous range of possible outcomes. In spite of the fact that construction projects are typically discrete objects, aspects such as the expected traffic over a road may range from a certain minimum to a certain maximum. The amount of traffic within this range will affect the loading conditions and the settlements of a road. Consequently, maintenance schedules are influenced by these ranges, which is of interest while considering life cycle costs. Therefore, this type of scenario may be helpful in our projects

4   *Alternate futures* indicate scenarios that result into a few *discrete* outcomes, contrary to the *range* of outcomes of the previous scenario type. This will feel as coming home to many of us. With this type of analysis we can for instance compare the scenario of a bridge with that of a tunnel.

Because it is all about trying to deal with the future, van Oirschot (2003) refers to scenario analysis as some kind of future management. However, scenario analysis has nothing to do with any crystal ball. This is not about *predicting* the future, rather *anticipating* on future developments, by using our experiences and imagination to create a desired and successful construction project. In the words of David Blockley and Patrick Godfrey (2000): 'Evidence of progress to success comes from past performance, present performance and possible future performances (scenarios)'.

By the presented types of scenario analysis it will be possible to reveal a wealth of foreseeable risks and opportunities for our projects. Altabba et al. (2004) advise applying scenario analysis for a detailed description of the entire project. In particular, the entire *process*, in other words the six distinguished project phases through which the project is realized, is important to consider in their view. Baya et al. (1997) suggest considering the extremes of a *dream scenario* and a *doom scenario*. It is likely that reality will arise somewhere in between these extremes.

As a result of a scenario analysis, each of the identified (ground) risks of each of the scenarios can be *classified* by the next GeoQ step. A practical ground risk classification tool will be presented in Chapter 9 about the pre-design phase. According to the GeoQ process, after risk classification the steps of risk remediation, risk evaluation and risk mobilization follow. In most cases these steps are not yet relevant to ground risk management in the feasibility phase, as first the project feasibility has to be decided upon.

One of the results of the GeoQ process in the feasibility phase is a clear definition of the required ground data for the next pre-design phase, further to verify and classify the identified ground-related risks. Without the GeoQ process, the definition of the required ground investigation will be mainly based on rather subjective experiences and judgement of the involved professionals. By using the GeoQ process, the knowledge of the identified foreseeable ground risk serves as a foundation for the definition of an effective ground investigation programme, in which the identified ground-related risks get the attention they need.

Finally, scenario analysis may also be of great use to *classify* the identified risks and to define appropriate risk remediation measures. In addition, it has been widely proven successful to develop so-called *fall-back scenarios* for the construction phase. The following two case studies aim to give some insight into the practical application of the GeoQ ground risk management process in the feasibility phase.

Case studies

This section presents two case studies of projects in the *feasibility* phase, and aims to demonstrate the appraisal of project feasibility from a ground risk management perspective. GeoQ was in its full development at the time these cases took place in practice. For this reason, not all presently available GeoQ steps and approaches could be fully applied. The first case was in fact the very first attempt to apply the GeoQ process, in order to evaluate an existing project with regard to the decision-making processes in relation to ground-related risk. The second case already reveals clear contours of the presently available GeoQ process steps and tools

Risk based decision-making for a light-rail project

The development of the GeoQ process started in 2001 with a risk-based evaluation of an infrastructure project. The trigger was the need of a project director, representing the client. He and his project team wanted to be able to decide for themselves about the feasibility of geotechnical options for their project, rather than relying on one single solution by a geotechnical consultant. Therefore, he suggested another approach of geotechnical reporting, based on discrete scenarios and risk analyses. He wanted to have a number of geotechnical options for his most important project decisions, together with an indication of the associated risks. In order to verify the feasibility of this innovative demand, it was decided to evaluate one of his projects, which was in the final construction phase at that time. GeoDelft performed the evaluation, with the author as project manager, and it turned to serve as a foundation for the development of a ground risk management concept, that later was named GeoQ. The main results of this evaluation were published in The Netherlands (van Staveren and Bolijn, 2003) and the evaluation is concisely presented and discussed below. The case aims to demonstrate how careful consideration of the ground aspects contributes to more transparent, rationally motivated and even better project decisions.

Let us begin with the first GeoQ step of *gathering* project *information*. The so-called Zuidtangent is aimed to become the future light-rail connection between Schiphol International Airport and the city of Haarlem, located west of the Dutch capital, Amsterdam. This densely populated area, in which, in fact, every square metre has been used, is a typical *greyfield* site. The routing of such an infrastructural project is determined by all kinds of town and country planning issues, including local political factors. Ground conditions are not a relevant factor for the alignment and should be considered as a given to deal with. In this part of The Netherlands, these ground conditions typically consist of some 10 to 15 m soft deposits, mainly clay and peat, underlain by Pleistocene sand layers. Ground water levels are high, often some 0.5 m below ground level. Furthermore, groundwater overpressures are locally present in the Pleistocene sand layers, due to artificial groundwater management of the polder areas.

For the purpose of the project evaluation, the client's project team identified, with their hindsight knowledge, the top ten most important project decisions they had to make. This exercise can be considered as some sort of backward risk analysis. Six out of ten decisions typically had an *alternate futures scenario* character. Each of these included two options, such as: should a canal crossing be a bridge or a tunnel? The effects of each of these scenarios on the project success factors planning, costs, quality and reputation were assessed in a qualitative way. Obviously, during the actual project process, these effects were *uncertainties* or *risks* of not meeting the project's objectives. Table 8.1 presents, in an arbitrary order, these six sets of scenario and their appraised impact on the project success factors.

Crossing 5 of the future light-rail passes an existing road, while Crossing 6 passes a canal. Table 8.1 presents the decisive factors in **bold**. In two of the six scenario sets costs were dominant in the selection of the most suitable scenario. Quality and reputation dominated each in two of the remaining scenarios sets. If we connect these scenarios sets to ground conditions, we arrive at the core

Table 8.1: Scenarios and their impact on project success factors

| Alternate future scenarios | | Effects on project's success factors | | | |
Set	Description	Planning	Costs	Quality	Reputation
1	Crossing 5: below or at surface level?	Large	Large	Large	**Large**
2	Crossing 6: tunnel or bridge?	Small	Large	Large	**Large**
3	Crossing 6: in situ or prefab construction?	Small	**Large**	Small	Small
4	Poelpolder area: below or at surface?	Large	**Large**	Large	Large
5	Part 5 to 6: concrete or earthworks foundation?	Large	Large	**Large**	Large
6	Entire route: concrete or asphalt pavement?	Small	Large	**Large**	Small

of the evaluation. For each scenario set the following three questions have been answered by the client's project team:

1   Was ground information *available* for selecting the scenarios?

2   Was ground information *critical* for the scenarios?

3   Was ground information actually *considered* for selecting the scenarios?

The ground information was of both factual and interpretative type and consisted mainly of the results of cone penetration tests (CPTs), the alternative for borings in the Dutch soft soils. Results of some preliminary geotechnical calculations were also available. Table 8.2 demonstrates the answers to the three questions for each of the six sets of scenario.

According to Table 8.2, ground information was available for each scenario set. In five out of six scenario sets, ground information was assessed as critical. However, in three out of six scenario sets, which was 50 per cent of the cases, the available and critical information was remarkably *not* considered in the decision

Table 8.2: Scenarios and ground information

| Alternate future scenarios | | Ground information | | |
Set	Description	Available?	Critical?	Considered?
1	Crossing 5: below or at surface level?	Yes	Yes	**No**
2	Crossing 6: tunnel or bridge?	Yes	Yes	**No**
3	Crossing 6: in situ or prefab construction?	Yes	Yes	Yes
4	Poelpolder area: below or at surface?	Yes	Yes	**No**
5	Part 5 to 6: concrete or earthworks foundation?	Yes	Yes	Yes
6	Entire route: concrete or asphalt pavement?	Yes	No	No

Table 8.3: Effects of two ground-related scenarios on the project's success factors

Scenario set number 5 Two light-rail foundation options	Planning (months)	Effects on the project's success factors		
		Costs (million euros)	Quality (remaining settlements in m)	Reputation (additional construction time in months)
Settlement-free concrete slab	2.5	1.14	0.00	0
Earth embankment, including a geotextile	9.0	0.95	0.04	3

process (presented in **bold** in Table 8.2). This is a remarkable result and supports the use of the GeoQ process approach, to guarantee that both available and critical ground information are considered when deciding about the most favourable scenarios for the following project phases.

For those readers interested, Table 8.3 presents the effects of two options for scenario set five on the project success factors planning, costs, quality and reputation. This illustrates how the effects of ground-related scenarios can be made explicit and expressed in terms of the project success factors.

In spite of its rather evaluative character, the presented approach should be well applicable in the feasibility phase of any project, where the relevant ground data may contribute to making the best possible decisions about the go–no-go of the particular project.

Risk driven planning of an urban development project

This second case about the feasibility phase describes a risk-driven domestic development planning of the Almere Pampus project, located in a polder below sealevel, about 30 km east of the city of Amsterdam. An area of 8 million m² will be developed to an urban area for domestic housing, including the necessary infrastructure of roads and sewerage systems. It also includes recreational areas with some lakes and parks. In spite of the fixed location of the entire project, the planning of all structures and typical functional locations *within* the area was still to be decided upon. This situation classifies the project as of *greenfield* type in which ground conditions play a dominant role in the actual planning of the entire domestic area.

The local government of the city of Almere had already provided three layouts or scenarios for the set-up of the project area. However, almost as usual, ground conditions were not yet involved. The Dutch Almere region is notorious for its heterogeneous ground conditions resulting from a dynamic geological history. Soft clay, peat and densely packed sandy deposits vary highly at very short distances of even a few metres. Because of the locally highly unfavourable and soft ground conditions the client requested a risk-driven assessment for the optimum layout of the area. The objective of this exercise was to save a maximum amount of money for site-preparation activities and maintenance during the entire lifetime of the domestic area. The results should support decisions concerning the financial feasibility of the project.

The consultant decided to provide an additional scenario for the layout of the entire urban development project, in which different structures and functions of the area were matched with local ground conditions. Recreational areas, with a lot of water and parks, should ideally be situated at locations with the worst soft soil conditions, while roads and houses should be located in areas with the most sound and favourable ground conditions (Hounjet, 2005). Obviously, the ground conditions would not rule the entire layout of the project, however, major potential cost savings were anticipated by a smart use of the given local ground conditions. Reverse situations, with the most sensitive structures located at the worst spots, from a ground conditions point of view, has been shown to be a reality, at least in The Netherlands.

Considering the first GeoQ step of *gathering information*, quite a lot of ground information proved to be already available:

- A geological map with a subdivision into favourable and unfavourable areas for foundations

- The thickness of the Holocene ground layers, consisting of mainly soft deposits

- The level of the top of the underlying Pleistocene sand layers, which form normally a sound foundation layer

- Expected settlements in case of 1 m fill for site preparation

- Expected foundation levels for the light structures.

This information was presented on five separate maps and based on this information a number of scenarios for the entire area were developed. Three main functions for the new urban have been distinguished:

1   Domestic housing

2   Roads and sewerage systems

3   Recreational areas with a lot of water and parks.

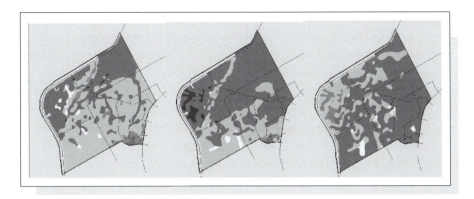

Figure 8.6: Risk maps for three functions of an urban development area (© with permission of GeoDelft).

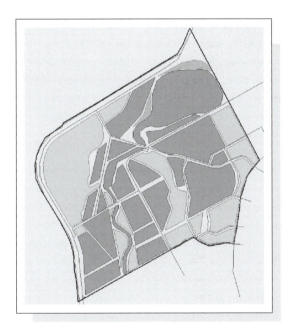

Figure 8.7: Ground risk remediation: matching functional areas with expected ground conditions (© with permission of GeoDelft).

Next the two GeoQ steps of *risk identification* and *risk classification* were taken. For each of the three identified functions a separate map was developed, which indicated areas with favourable and unfavourable conditions for that particular function. The favourable areas involve a low risk for additional costs, extra construction time and reduced quality, because these areas were suitable with regard to the proposed functions. The unfavourable areas present a relatively high risk, as these areas are assessed as less suitable for the particular function. Figure 8.6 presents these three maps, in which the relatively dark shaded parts represent the unfavourable areas. The relatively light shaded parts in Figure 8.6 represent the more suitable locations for the particular function.

As a measure of *risk remediation*, GeoQ step four, these three maps were combined into one map with the most suitable layout of the development area, from a ground conditions perspective. This map is presented in Figure 8.7 in which different shades identify different functions.

As an agreed rule within the project team, the most sensitive function was located in the areas with the most favourable ground conditions. Therefore, the sewer systems, which are relatively sensitive to differential settlements, were located as much as possible in areas with the least expected settlements.

In the next GeoQ step of *risk evaluation*, the map of Figure 8.7 was compared with the three initial existing layouts, which has omitted any ground condition risk. It became clear that the newly proposed layout would result in a reduction of 1000 m^3 sand fill per 10 000 m^2. This will significantly reduce both costs and site preparation time. The speculative settlement risk, with settlements either larger (negative) or smaller (positive) than anticipated, turned out to be positive in this risk assessment and compared with the initial situation. A foreseeable and speculative risk has been transformed into an attractive opportunity. If the maintenance costs during the lifetime of the project are included, then the risk-driven alternative layout results in 20 per cent lower maintenance costs, compared with the most unfavourable initial scenario. The total cost savings for the municipality of Almere, for a maintenance period of 50 years, are calculated at 64 million euros (Pereboom et al., 2005).

The final GeoQ step, *mobilization* of all retrieved ground risk information to the next project phase of pre-design, could easily be made, because of the ready available risk maps. These maps were all digitally available in colour (unlike the prints in this book), which make them very easy to use by all kinds of project participants in the following project phases. In addition, new ground information, as a result of detailed ground investigations for roads and houses can be easily added to the maps, by today's available Geographic Information Systems (GIS) and related software packages.

To conclude, the application of the GeoQ ground risk management process in an early phase of the urban development project revealed an opportunity for a major cost saving. Not only did the negative side of ground conditions and risk became visible, opportunities also surfaced. This case supports the statement that ground risk management, in an early phase of a construction, may contribute to both risk reduction and project optimization.

Summary

This chapter demonstrated the application of the GeoQ process in the feasibility phase of a construction project. It focused on the first two GeoQ steps of *gathering information* and *risk identification*. Three approaches to support ground risk management in this earliest project phase have been introduced and discussed: project site classification in terms of greenfield, brownfield or greyfield sites, how to do more with a minimum of ground information from a risk management perspective, and the concept of scenario analysis, of which the range of futures and alternate future types tend to be most useful for ground engineering and construction activities.

These GeoQ approaches appear to be viable for ground risk management in the feasibility phase. As the two cases have demonstrated, the approaches do not only facilitate risk identification at an early stage, they may very well reveal attractive opportunities as well. Both support the project's go–no-go decision at the end of the feasibility phase.

9   GeoQ in the pre-design phase

Introduction

The previous chapter covered the very first phase of any construction project, this chapter continues with the next phase in line: the pre-design phase. Figure 9.1 positions this phase within in the six phases of the GeoQ process.

During pre-design, several engineering solutions will be considered and compared, such as a cut-and-cover or bored tunnel? GeoQ ground risk management facilitates this process in view of the anticipated ground-related risks and opportunities. The optimum design solution will be selected at the end of the pre-design phase, based on the project specifications, the risk tolerance of the parties involved and the foreseen ground conditions. For instance, it will be decided to continue the tunnel project with a bored tunnel.

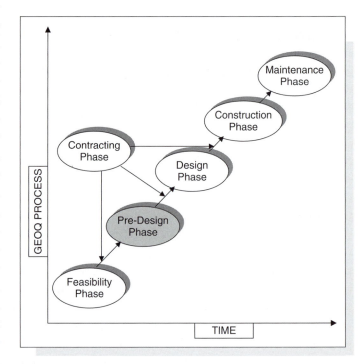

Figure 9.1: The pre-design phase within the six GeoQ phases.

This chapter starts with the introduction and discussion of three approaches for supporting ground-related risk management during pre-design. These methods are specifically related to the three GeoQ steps of risk identification, risk classification and risk remediation. The first approach is a team-based risk identification and classification procedure. This approach can be facilitated by the so-called Electronic Board Room, which is an ICT-supported tool. The risk identification and classification procedure results in a risk profile of the project.

The second ground risk management approach explores the relationship between costs of ground investigations and risk. It aims to balance the established project risk profile with the optimum scope of ground investigations. An adequate ground investigation is a way to risk remediation and, based on its results, we should be able to select the most suitable pre-design solution.

The third ground risk management approach is closely related to the second one and presents a number of considerations for a risk-driven ground investigation. This method describes how to balance ground investigations with risk. For a ground investigation in the pre-design phase, the type, quantity and quality are defined in a risk-driven way, while matching the project's risk profile. Two case studies are presented, in which the presented ground risk management methods have been successfully applied. As usual, the chapter finishes with a summary.

Ground risk management during pre-design

Team-based risk identification and classification

The first ground-related risk management approach, which is particularly useful in the pre-design phase, concerns a practical way for identifying and classifying risk. We will explore the risk brainstorm session, in which groups of experts, based upon their mixture of expertise, experience and perceptions, identify and classify risks.

Preparation of the risk session

Like other brainstorm sessions, the team-based risk session needs to be carefully prepared to become successful. Its objectives, risk identification and classification structure, proposed participants, and programme require due attention. Preferably the project's key success factors, with regard to costs, planning, safety, quality and reputation are already expressed in measurable targets. These are the benchmarks for the risk identification and classification process, because the overall project risk is not meeting these objectives.

The risk structure also needs to be considered before starting any risk identification. For instance, risk of technical and non-technical origin can be distinguished and it is possible to subdivide ground-related risks in geotechnical, geohydro-logical, geoenvironmental and man-made structure risk groups. Furthermore, a project-specific risk classification system needs to be selected. Typically, the semi-quantitative method is applied in team-based risk sessions.

The participants for the risk brainstorm session have to be selected with due care. In the pre-design phase of a project, the risk session participants probably have a multidisciplinary character and a variety of backgrounds, expertise and experience, to identify and classify the widest range of foreseeable risk. During the later design and construction phases similar sessions with more mono-disciplinary experts are able to focus on the remediation of specific ground-related risks.

Risk identification

Two main streams of team-based risk identification can be distinguished:

1   Building further on already identified risks

2   Starting with an entirely fresh risk identification process.

The main advantages of the first approach are efficiency and focus. Many identi-fied and already structured risks, possibly retrieved from preparatory individual interviews, completed risk questionnaires, and risk registers and databases, serve as the foundation for the group session. However, participants in the risk session are already influenced and biased by the existing risk information, which may severely reduce their out-of-the-box thinking capabilities during the session. The latter is required to bring as much as possible risk from the unforeseen to the foreseen risk space. For this reason, the second way may be most suitable in cer-tain project occasions. It is up to the risk coordinator in charge to select to most suitable way for his or her particular project.

Once identified, the risks need to be prepared for their classification as major or minor risks, which requires special attention to the people factor. According to Blockley and Godfrey (2000), the human mind appears to be more effective at addressing success than failure. Keizer et al. (2001) bring up the prospects theory, which shows negative framing of risks to result in more positive risk perceptions than positive framing. In confrontation with negative statements we intend to respond by considering that 'it should not be as bad as that'. Positive statements, however, seem to trigger our awareness of possible pitfalls, in terms like 'I am not so sure about this statement'. It motivates Keizer et al. (2001) to translate identified risk towards so-called positive formulated statements. A risk statement like 'differential settlements cause structural damage' can be positively formulated in 'differential settlements will be limited in order to avoid structural damage'. This type of risk reframing probably needs some more preparation time and guidance.

Risk classification

It is possible to express the seriousness of a risk in the probability of occurrence and the likely effects or consequences, if the risk indeed occurs. Therefore, risks are often classified in terms of probability and effects. Reframing risk becomes some sort of challenge, because it appears rather difficult to classify positive risk statements in the conventional way of probabilities of occurrence and effects. We still want to assess probability and effects if the positive statement of 'differential settlements will be limited in order to avoid structural damage' is not reached. In other words, we still have to appraise the probability and the effect of the occurrence of the negatively framed statement 'differential settlements cause structural damage'. For this reason, I tend to choose the conventional negatively framed way of classifying risk. A team with a variety of risk perceptions should be able to prevent over-optimistic risk classification.

Figure 9.2 presents a proven matrix for semi-qualitative or semi-quantitative risk classification. The terms 'semi-qualitative' and 'semi-quantitative' risk have here the same meaning and are widely used in an intermingled way.

Figure 9.2 distinguishes six risk classification criteria and these are applicable to each risk. Three of them relate to the risk probability of occurrence. The remaining three criteria classify the risk effect. The knowledge about the risk concerns, for instance, expertise about in-situ transport mechanisms of polluted groundwater in an aquifer. The period of occurrence is an indicator for the time frame in which the risk can take place. A deformation risk in a building pit is largely limited to the time required for the excavation and installation of any struts. A differential settlement risk, however, typically continues over a long period of time and often proceeds after completion of the project's construction. Similarly, the effect-criteria of time, costs, as well as safety, quality and reputation can be scored according to the criteria presented in Figure 9.2.

	RISK CLASSIFICATION CRITERIA	
RISK PROBABILITY	*RISK SCORE = 1* *(criterion with low significance)*	*RISK SCORE = 2* *(criterion with high significance)*
Knowledge	a lot of knowledge available	limited to no knowledge available
Period of occurence	less than 1 month	more than 1 month
Way of occurence	slowly and gradually	fast and sudden
RISK EFFECT		
Time (planning)	less than 1 month delay	more than 1 month delay
Costs (budget)	less than 50,000 Euro extra costs	more than 50,000 Euro extra costs
Safety, quality and reputation	some reduction is acceptable	any reduction is not acceptable

Figure 9.2: A semi-qualitative risk classification matrix.

This type of risk classification needs to be made fit-for-purpose for the project of consideration. The criteria for the risks scores have to be adapted to the project's objectives and character. In addition, everyone is free to add other criteria or even to delete criteria, again depending on the type of project and the degree of risk awareness of the team. For instance, it is possible to add the ability of the project team to influence the identified risks, expressed in risk control, as a third main criterion besides risk probability and risk effect, as suggested by Keizer et al. (2001). Also the risks scores can be easily revised, by adding one, two or even more risk scores.

The example shown in Figure 9.2 refers to a rather small project, which is reflected in the rather small extra costs criterion of 50 000 euros. For a larger project these extra costs can be extended to 0.5 million euros or even 5 million euros.

Figure 9.3 shows ten identified and scored ground-related risks and has been retrieved from a team-based risk session with a contractor for a bridge project.

Figure 9.2 showed a risk classification matrix with a simple risk score, limited to one and two. Each of the ten risks as shown in Figure 9.3 has been scored one, two, or three for each classification criterion, where a one represents a low score and three refers to a high score. As an agreed rule, every participant of the risk session should only score the classification criteria of which he or she has a judged opinion, based on expertise and experience. The scores shown in Figure 9.3 are the average values of the scores of the individual participants on each criterion.

It is useful to analyse the number of scores for each individual risk criterion and to calculate the standard deviations as well, because this provides viable information about the risk. For instance, if only one participant scored the costs criterion for risk number one, the presence of archeological remains, then a serious indication of lacking knowledge within the team has become clear. The sum of the risk factors that classify probability (knowledge, period and way of occurrence) result in the probability risk score (Ptot). In theory, the maximum risk score for both probability and effect is nine, the sum of the maximum score of three for each individual criterion. The minimum score is three, which is the sum of the minimum score of one for each individual criterion. The same applies for the risk criteria that classify the total risk effect, which results in the effect risk score (Etot). If required, Ptot and Etot can be summed or multiplied to arrive at a single risk score number of each classified risk.

Results of the risk session

The results of the risk session are usually plotted in a risk matrix. The matrix has two dimensions and presents the total scores on risk probability and risk effect for each classified risk. Each classified risk of Figure 9.3 has been plotted in Figure 9.4 and the risk numbering of Figure 9.3 corresponds with Figure 9.4.

Risk No.	Ground-related risk description	Probability Risk Score			Effect Risk Score			Total Risk Score	
		K	P	W	T	C	S	Ptot	Etot
1	Presence of archeological remains	2.6	1.7	2.6	2.4	1.1	1.3	6.9	4.8
2	Horizontal and vertical deformations	1.4	1.9	1.6	2.0	2.3	2.7	4.9	7.0
3	Geotechnical parameters worse than assumed	1.5	1.9	1.6	2.4	2.6	2.3	5.0	7.3
4	Presence of bombs and grenates in the ground	2.3	1.9	2.6	1.6	1.4	1.3	6.8	4.3
5	Sand bearing layer deeper than assumed	1.8	1.4	2.5	1.8	2.0	2.0	5.7	5.8
6	Geotechnical model is not correct	1.8	1.8	1.5	2.2	2.6	2.8	5.1	7.6
7	Factual ground data is not correct	2.0	2.1	1.8	2.5	3.0	2.4	5.9	7.9
8	Assumed loading conditions are not correct	1.2	1.8	1.6	2.2	2.6	2.3	4.6	7.1
9	Heterogeneous ground conditions	2.0	1.5	1.4	1.8	1.8	1.6	4.9	7.3
10	Presence of polluted ground or groundwater	2.4	2.3	2.1	2.0	2.2	2.0	6.8	6.2

Legend:

K = Knowledge about risk (influence on risk probability)
P = Period of occurence (influence on risk probability)
W = Way of occurence (influence on risk probability)
T = Time refers to planning (influence on risk effect)
C = Costs refers to budget (influence on risk effect)
S = Safety, Quality and Reputation (influence on risk effect)
Ptot = sum of the scores of K, P and W
Etot = sum of the scores of T, S, C.

Figure 9.3: A semi-qualitative score of ten ground-related risks.

Evidently, the more the risk is located in the upper right corner of the risk matrix, the more severe the classified risk. Figure 9.4 identifies three risk zones by two diagonal lines. The zone below the line that connects the total risk scores of six can be considered as a low risk zone. The zone above the line connecting the scores of nine is typically a high risk zone. The zone in between presents the intermediate risks. Obviously, the risk team may decide about the most appropriate risk zonation for their project.

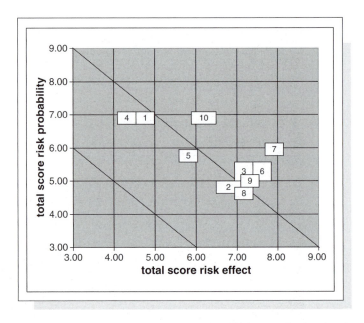

Figure 9.4: A risk matrix with ten classified ground-related risks.

Any risk remediation measure, by risk cause reduction, risk effect reduction or the combination of both, will become visible in an updated version of the risk matrix. Typically, the resulting residual risk will move in the lower-left direction of Figure 9.4. A repetition of this risk classification process in a later project phase will reveal these effects.

ICT-supported risk identification and classification

The Electronic Board Room (EBR) is an ICT-facilitated method for team-based risk brainstorming. It proves to be a fast and effective method for the identification and classification of project risks by the approach as discussed just before.

During an EBR session, 10 to 15 (laptop) personal computers are connected to a software package. Behind each computer one or two professionals identify risks, while they follow the results of the other EBR participants, real-time, on their screens. This may trigger them to identify other and new foreseeable risks, which is the brainstorming part of the session. Figure 9.5 shows a typical setting of an EBR session.

Usually, the EBR system works anonymously, which means that the participants do not know who has identified which risk. This approach minimizes the unwanted group dynamics of conventional brainstorm sessions. You have probably also encountered those good-intended but rather loud and convincing voices who overrule the other participants and adversely dominate the brainstorm procedure.

Figure 9.5: A typical setting of EBR-facilitated and team-based risk session (© with permission of GeoDelft).

Besides risk identification, the EBR facilitates risk classification as well. It is easy to apply the described semi-qualitative method. Table 9.1 presents a typical programme for an EBR session that proved to be successful in practice for a large number of projects. It demonstrates the feasibility of performing a risk identification and classification in just half a day.

Table 9.1: Programme for an EBR-facilitated and team-based risk session

Time (hours)	Activity
09.00–09.30	Introduction, including project and session objectives
09.30–10.15	Risk identification
10.15–10.30	Coffee and tea break
10.30–10.45	Preparation of risk classification
10.45–11.30	Risk classification
11.30–12.00	Presentation of results in risk matrices
12.00–13.00	Joint lunch, agreements on next steps and closure

Obviously, depending on the characteristics and objectives of the project of concern, alternative programmes may be preferred and established. Anyhow, each session will start with a general introduction about the session's objectives and the risk assessment software to be used. This software is user-friendly and can normally be operated by every participant just after a concise test run.

Then the risk identification takes place, either by building on existing risk files or by starting from zero. After a check of the identified risks, with regard to their appropriate structuring in pre-defined risk groups, the risk classification system will be introduced, to be followed by the risk classification. Two major advantages of the ICT-facilitated risk identification and classification are efficiency and the digitally available results, as presented in Figures 9.3 and 9.4. Other formats for risk presentations can be easily derived when demanded by the project circumstances. This risk information serves obviously as a sound basis for the project's risk register.

After the team-based risk session

Based on the results of the team-based risk session, the next GeoQ steps of risk remediaton, risk evaluation and risk mobilization can be made. The steps are extensively described in Chapter 10 through to Chapter 13. In addition, each risk needs an explicit allocation to one of the project participants, such as the client, contractor and perhaps even the engineer, as discussed in Chapter 11.

The definition and performance of a ground investigation is one of the main risk remediation measures after the team-based risk session in the pre-design phase. Besides the sort of site – greenfield, brownfield or greyfield – the scope of the ground investigation should be highly driven by the results of the project risk profile that resulted from the risk identification and classification process. In other words, which risks should be remediated by the additional knowledge of a ground investigation? This demands balancing risk profiles and ground investigations.

Balancing risk profiles and ground investigations

The dream of optimum ground investigations

Anyone involved in design and construction understands that any form of ground investigation is required in order to arrive at an effective design and construction. Nothing new so far. Every ground investigation does not just need proper planning, it calls also for creativity, a critical attitude and a risk-driven approach. Typical questions to be asked are of the type: 'Is a standard investigation sufficient or is it worthwhile performing some additional tests?' An appropriate answer to this type of question is highly risk profile dependent.

A wish I regularly encounter, expressed by both ground engineering experts and lay people who have to manage ground-related aspects in their projects, concerns the suitable scope of the ground investigation. A question that echoes in each and every construction project, at a certain moment, is: 'How much ground investigation do we have to do?'. The hidden wish underlying this question is a clear and uniform answer, that says: 'With so many boreholes and those laboratory tests we have exactly the right ground investigation scope for our project'. Unfortunately, reality is not that nice and such an answer is simply non-existent. The optimum ground investigation depends on the earlier indicated variety of project specific factors. Perhaps even more important, the right scope of ground investigation is not an objective fact. Here we revisit the soft systems and people factor within any construction project. The most appropriate ground investigation is particularly dependent on the subjective perceptions of the ground-related project risks. Therefore, the best thinkable ground investigation varies (widely) with probably most of the individual experts within the project. While interpreting ground data, who does not recognize sighs of ground engineers, like: 'How is it possible that these guys did not take a sample of this very important ground layer'. This is due to the differences in risk perception between these individual professionals, the effect of team dynamics on risk attitude and not forgetting the risk tolerance of our clients. Therefore, this entire book will not provide one formula with an absolute validity for the optimum ground investigation. That dream has gone.

From dream to reality – the relative costs of ground investigations

However, what we can do is highlight the relationship between ground investigations and the project risk profile. Based on the first three GeoQ steps of gathering information, risk identification and risk classification, we have to remediate the most important pre-design risks in some way or another. Normally, we need additional ground information to be able to take suitable risk remediation measures. We should balance the costs of these ground investigations with the seriousness of the risks we need to remediate. Figure 9.6 presents the balance we are looking for.

Figure 9.6 presents two trends with, in my view, only a theoretical validity. The first trend relates a ground risk decrease to an increase in the scope of ground investigations. In other words, if more ground data (the knowledge factor) were available, then the degree of ground-related risk would reduce. This trend agrees with our gut-feeling but is usually hard to prove by factual data. We will recognize the second trend from our practice: an increase in the scope of ground investigations results in an increase in the ground investigation costs. Our main challenge is to balance ground investigation costs with an appropriate reduction of ground risk. The resulting ground risk profile should be in due balance with the initial costs of the required ground investigation. Deliberately, Figure 9.6 presents this optimum as in a cloud. In my vision, this optimum cannot be reached in an objective way because the inherent uncertainty of the cost effects of the residual risks can never be compared with the certainty of the ground investigation costs within a certain scope of work. We remain at best guesses, unfortunately.

However, the costs of ground investigations and related activities are relatively low when expressed as a percentage of the total project costs. According to Smith (1996), costs of ground investigations are typically less than 1 per cent of the construction costs. Blyth and de Freitas (1984) assess the costs of ground investigations to be between 0.5 and 1.0 per cent of the project costs, but this should not be taken as a rule. Ground investigation costs between 0.2 and 0.5 per cent are typically considered as adequate in the construction industry (Knill, 2003), while

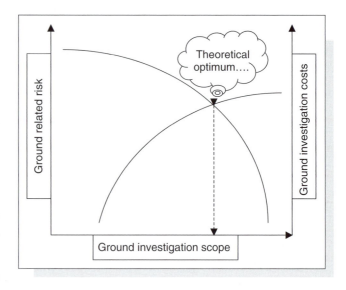

Figure 9.6: The theory of balancing costs of ground investigation with risk.

Brandl (2004) estimates ground investigation costs to be between 0.1 and 1 per cent of the project costs. Indeed, the literature presents a variety of percentages, which have in common that they are typically (far) below 1 per cent of the total project costs.

In addition to this wide ground investigation costs scatter, Brandl (2004) adds that 'money cannot buy risk free ground', even if the usual percentages of ground investigation costs were to be multiplied by a factor of 5 to 20. As far back as 1984, the US Subcommittee on Geotechnical Site Investigations and the US National Committee in Tunneling Technology recommended increasing the expenditure on ground investigations to an average of 3 per cent of the project costs in order to arrive at better overall project results, at least for typical underground constructions (US Subcommittee on Geotechnical Site Investigations, 1984). Their evaluation of 22 case studies of underground construction projects demonstrates a clear connection between expenditure on ground investigations and the deviation of the final project costs from the initially estimated project costs. Projects with ground investigation costs up to 1 per cent of the estimated total project costs, showed deviations in final project costs from −40 to +70 per cent, compared to the estimated total project costs. With an expenditure on ground investigations of typically more than 2 per cent of the estimated total costs, these deviations stayed within a margin of 10 per cent. Clayton (2001) presents the results of an investigation by Mott McDonald and Soil Mechanics Ltd (1994) for more than 50 UK highway projects. The evaluation of these projects confirms the relationship between construction cost overruns and the degree of investment in ground investigations. The US and UK research also confirm the relationship between the degree of ground-related risk and the scope of ground investigations, as presented in Figure 9.6.

If we recall the facts of ground-related problems easily adding 5 per cent, 30 per cent, 50 per cent or even more than 100 per cent to the initial project price (Clayton, 2001), then there is a huge potential added value of adequate ground investigations. However, the attitude towards ground investigations is sometimes not yet favourable to bring this potential into practice.

Towards risk driven cost awareness

As presented in Chapter 2, it is estimated that a significant portion of all failure costs in the construction industry are directly or indirectly connected to uncertainties related to the ground conditions. At least in The Netherlands and, for instance, also in Norway, the USA, China and Japan, there is an unfavourable history and an ongoing habit to buy ground investigations occasionally on the cheapest price criterion only. The client or client's engineer provides a 'shopping list' with a number of in-situ and laboratory tests, often with minor or even no context of the project requirements. The bidder with the lowest price wins the contract, and apparently everybody is happy. As demonstrated by the previous studies and examples, these types of ground investigations, without a well-defined risk driven strategy and not seldom of a doubtful quality, are by far not the most cost effective at the end of the project. It is even more remarkable when we look at ground investigations from a medical point of view. Box 9.1 presents a comparison of ground investigations with medical investigations.

Ideally, the choice of the type of 'geotechnical hospital' to define the most effective and efficient scope of ground investigations should be based on the results of the ground risk identification and classification. A number of standard penetration tests in the field ('the family doctor') are usually sufficient in rather small-scale and simple projects in the early design phase. For larger and more complex projects it is often worthwhile performing some undisturbed sampling and laboratory tests ('the local hospital'). In the case of serious risks or a low risk tolerance of the parties involved, innovative site investigation techniques ('the specialist hospital') may provide important added value, possibly even in the pre-design phase.

We are able to act in a risk-driven way in order to attempt to optimize our ground investigations. Indeed, an early and risk-driven approach to ground investigations and consultancy services may be initially more expensive than the

---

Box 9.1 Ground investigations from a medical perspective

Let us explore the actual ground investigation approach from a medical perspective. If you visit your family doctor with a broken leg, you cannot expect
that he or she is able to provide you with the necessary treatment. You will be
sent to a local hospital where your leg will be set in splints. In the case of a
complicated fracture, you will probably be treated in a specialist hospital. For a
whole range of other health problems, though, your family doctor is certainly
the right person to approach. We accept this situation as being completely
normal.

The fact that you cannot consult your family doctor for each and every problem is, however, far less obvious when it comes to geotechnical engineering
in the construction industry. For example, to manage geotechnical problems,
such as the stability of a construction pit which is located just 1 m from a
historical building, it is often the 'family doctor' who is approached in such
cases, since this often seems to be the cheapest option. In reality though, the
complex ground behaviour and the associated risks often require a 'local'
or even a 'specialist hospital', in order to minimize the total final project
costs.

---

conventional approach. It may, however, deliver tangible results, as indicated by
the previous research examples. Ground risks can be managed more effectively.
The potential savings are generally far higher than any initial additional costs that
may be incurred by a sound ground investigation, when we remember the presented percentages on ground investigation investments in relation to the anticipated
deviations in ultimate project costs. In addition, this type of risk-driven ground
investigation does not necessarily always mean more expensive investigations.
When extending a motorway section in The Netherlands, an alternative site investigation programme resulted in cost savings of 35 per cent in comparison with a
conventional approach. This was achieved by making use of existing ground data
(the results of information gathering) and a combination of cone penetration tests
(CPT) and relatively inexpensive electromagnetic measurements.

In conclusion, everyone involved in a project must realize that there is no
standard formula with regard to the optimum scope of the ground investigation.
Like in the medical sector, differentiation within the ground investigation sector is
highly necessary. Obviously, the selection of the ground investigation contractor
should not be based on the lowest price only. Apart from an attractive (low) price,
aspects such as quality and the type of investigations in relation to the risks and

project complexity have to be appraised. This seems to call for risk-driven ground investigations!

Towards risk-driven ground investigations

The main objective of the previous section was to create awareness about the fact that ground investigations should be balanced with the agreed risk profile of any project. This section explores how ground investigations can be balanced with the agreed risk profile in the pre-design phase. Here a major pitfall arises, the often approached mindset that a risk-driven approach is inherently difficult and complex, as for instance mentioned by Smith (1996) and Ho et al. (2000). Risk management does not necessarily imply difficult probabilistic calculations and a jungle of cause and effects diagrams. Also with regard to ground investigations, risk management starts with common sense and well-structured thinking about the main project goals to be achieved, the probability and effects of unfavourable events or risks and how to remediate these by proper site investigations. This type of analysis, which can be very brief for small projects and more extensive for large and complex projects, results almost fluently in the required ground investigations.

Therefore, by balancing ground investigations with the agreed risk profile of our projects, we should add a dimension to our conventional way of defining ground investigations. Likely the combination of codes of practice and the experiences of the involved experts determine the scope of the ground investigation, likely with the best intentions. By adding the third dimension of risk, from the results of the risk identification and classification processes, we should be able to answer effectively and efficiently the following three generic questions about ground investigations:

1 what type of ground investigation do we need?

2 what quantity of ground investigation do we need?

3 what quality of ground investigation do we need?

We should acknowledge here the statement of Clayton (2001), that about 85 per cent of all ground-related problems are directly related to the quality and extent of ground investigations. In the pre-design phase, the primary objective of a ground investigation is to arrive at a risk-driven conceptual model of the ground. Given the rather sketch-like character of the pre-design to be made, normally a preliminary, but risk-driven (!), type of ground investigation is sufficient. It should reveal the ground layering and different ground masses, together with the main geotechnical, geohydrological and geoenvironmental material properties of these ground layers and masses. Detailed ground investigations, to fill the white spots, the unknown areas of ground properties and behaviour that surfaced during the pre-design phase, will be performed in a later project phase.

Therefore, regarding the first question about the type of ground investigation techniques, usually the more conventional site investigation techniques, like (core) drilling with sampling and cone penetration testing, are appropriate. Laboratory testing involves normally the well-known index tests, such as sieve analyses, completed with testing of the deformation and the strength properties by oedometer testing, triaxial testing, unconfined compressive strength testing, and so on. In addition, geophysical techniques may add value, as well as aerial photographs and even satellite pictures. The added value of more specialist site investigation tools surfaces probably in the next phase, because then the really important risks are disclosed, including their risk driving parameters. Many specialist textbooks describe the numerous available drilling, sampling and cone penetration techniques. It is considered beyond the scope of this book to provide a detailed lists with all pros and cons of the existing site investigation tools. Specialist advice should be sought if we do not feel up to date about our actual ground investigation knowledge. This may turn out to be a very good investment in our project.

The second question is about the quantity of the site investigation. How many tests should we do with the selected techniques? Again, this question is very much dependent upon the site characteristics, the project characteristics and the agreed risk profile of the project. Therefore, the only answer I have is: it depends.

The third question about the quality is easier to answer: the quality should be according to the internationally widely available but still often neglected codes, standards and guidelines. We should be able to feel confident about the factual ground investigation results. Even if the factual data are correct and state-of-the-art in their presentation, the inherent fuzzy, random and incomplete character of ground conditions provide still ample challenges for a successful pre-design.

An increasing number of ground investigation contractors are ISO certificated. For the more large and complex projects a project specific quality programme may add value. The certification of field and laboratory investigation worldwide is in the very early development stage. I expect a real increase in certification, like in many other sectors, in the coming years, in particular if risk-driven ground investigations become more widely applied. Because with a risk-driven approach it is quite difficult to hide poor quality.

In my opinion, this risk-driven approach of the definition of the type, quantity and quality of ground investigations requires much more attention in day-to-day practice. The following case studies reveal some practical experiences with this risk-driven approach of ground investigations.

Case studies

In this section two case studies are presented concerning the pre-design of a project or parts of a project. The purpose of these cases is to demonstrate how the GeoQ process may work in practice in the pre-design phase of a project. The first case demonstrates how the GeoQ process facilitates the tender preparation for a quite large design and construct (D&C) project. The second case demonstrates how GeoQ can also contribute in the pre-design phase of a rather small project. It describes the process of ground risk management for a rather small directional drilling project.

As mentioned, GeoQ was in its full development during the cases studies. For that reason the steps and methods are not yet fully applied in these cases, with the hindsight knowledge and experience we now have gained so far.

Ground risk management for a tunnel project

As in many large cities worldwide, the city of Amsterdam also experiences its daily traffic jams. The capacity of one of the city's major tunnels, the Coentunnel, will be doubled by a second tunnel in an attempt to reduce some of these traffic jams. The Ministry of Transport and Public Works decided to apply the GeoQ approach in the pre-design phase of this design and construct (D&C) project, to support contract preparation for the tender. One of the main reasons for this client applying the GeoQ process was to prepare a transparent set of tender documents, in order to avoid, as much as possible, disputes about differing site conditions in the later phases of the project.

Let us start with the first GeoQ step of gathering information. The site of this second Coentunnel is typically a greyfield site, because of its location adjacent to the existing Coentunnel. There are no alternatives for the location of the second Coentunnel. The ground, with its geotechnical, geohydrological and geoenvironmental properties has to be considered as a given fact. Management of the interaction of the new tunnel with the existing tunnel will be one of the main challenges.

Because of the existing Coentunnel and motorways a lot of historical ground data are already available at the pre-design phase. The project was started a number of years ago and then became postponed. A lot of factual and interpretative geotechnical could be used, varying from data back in the 1960s to more recent data from the 1980s.

With the existing ground data and project specifications as a sound basis, the next two GeoQ phases of risk identification and risk classification were entered. For this purpose, experts from both the client and the engineer organized a team-based risk session, facilitated by the Electronic Board Room. During this session the foreseeable ground-related risks were identified and classified, according to the method as described before. Considering the objective of the preparation of D&C tender documents with a maximum of transparency about ground-related risks, the following six risk categories were distinguished:

1   Geotechnical risks

2   Geohydrological risks

3   Geoenvironmental risks

4   Risks caused by man-made structures in the ground

5   Contractual ground-related risks

6   Risks of unacceptable quality of ground data.

We may recognize the first four risk categories as the four main types of ground-related risks. The remaining two risk categories were added, because of the specific client's objective.

In total, 141 ground-related risks were identified, classified and stored in a ground-related digital risk register. Based on the risk register, adequate risk remediation measures were defined for the most important risks, which implies GeoQ process step 4. The two main ground risk remediation measures were:

•   The definition of an additional ground investigation

•   The preparation of a Geotechnical Baseline Report (GBR).

The identified and classified ground-related risks revealed the need for additional ground information, in order to be able to provide the tendering contractors an appropriate set of ground information. The definitions of the additional ground investigation programme were based on the type, quantity and quality of the existing ground data and the main ground-related risks. The costs of the ground investigation were carefully balanced with the actual ground risk profile of the project.

It became apparent that such an approach is still quite new. In particular, the experienced ground engineers in the project needed to transform their conventional way of engineering, with implicit ground risk management, towards the GeoQ process with explicit ground risk management. It required quite some change management competences of the project management in order to apply the GeoQ process as intended.

In spite of the project's early pre-design phase, the risk-driven approach resulted in the application of a few rather innovative and specialist site investigation methods. Serious risks of intolerable horizontal and vertical deformations of both the new and the existing tunnel were the main drivers for this advanced testing. It included in-situ measurement of the elastic ground properties by conepressiometer testing (CPM) and continuous ground profiling by using the so-called Consolitest, a geophysical tool. Furthermore, the compression moduli of the ground were carefully determined by so-called K_0-Constant Rate of Strain (CRS) laboratory testing (Pereboom et al., 2005). The ground engineering experts among the readers may recognize these tests. For the remaining readers the message is just that, even in an early project phase, advanced ground testing may be judged as cost-effective, when based on a risk analysis and serving as a risk remediation method.

The results of the additional ground investigation were very welcome to serve as an additional basis, together with the historical ground data, for the second ground risk remediation tool: the preparation of the Geotechnical Baseline Report (GBR). In this GBR each identified ground-related risk is clearly allocated to the contractor or the client. The GBR serves as a contractual document and is issued together with the other tender documents. Chapter 11 explores the GBR approach within the GeoQ process in detail.

The preparation of the GBR included the fifth GeoQ process step, the evaluation of the remaining risk profile. A digital database was built for the storage and use of all available ground information. It served also as tool to perform step 6 of the GeoQ process: mobilization of the ground risk data to the next project phase. In fact, the GBR and the database served as a combination in this respect. The GBR is part of the tender documents to the bidding contractors, together with a set of factual and interpretative ground data that can be easily retrieved from the database. Alternatively, the client can choose to provide the bidding contractors full on-line access to the database. Advantages are the reduction of transaction costs and full disclosure of all ground-related data to the bidding contractors.

One lesson learned should be added to this case. The contracting phase of this project was in full swing, during the writing of this book. In this tender phase, one of the bidding contractors lacked essential ground information, in spite of the careful and risk-driven preparation of all ground data in the tender documents. That particular contractor proposed an innovative tunnel design, which required very deep piling. As a consequence, in the view of that particular contractor, the ground investigation depth of about 30 m below ground level should have been extended to at least 50–60 m below ground level. This design scenario was clearly unforeseen by all experts involved in the pre-design phase. It is one of the consequences of a design and construct type of contract. With this lesson I want to confirm that GeoQ is not a panacea to avoid any ground-related problem. It may be a big step in the right direction, but as this case reveals, the ground, as well as the solutions to deal with it, are very difficult to predict.

Selecting the most suitable horizontal boring technique

This second GeoQ case study about the pre-design phase concerns a rather small project, as against the large tunnel project presented in the previous case. It is to demonstrate that the GeoQ process can also be applied for rather small projects in an effective way, without using sophisticated tools such as the Electronic Board Room for ground risk identification and classification. In this case the GeoQ approach was initiated by a contractor.

During the construction of a motorway in the south-eastern part of The Netherlands, it was necessary to relocate several existing pipelines. Complications were foreseen in that section of the project where the new motorway crosses a river. An important National Park-like nature conservation area is located in the vicinity of the river crossing and any excavation activities are forbidden by Dutch law. In the pre-design phase of this project the horizontal directional boring technique appeared to be an interesting option to relocate the pipelines below the river and to keep the protected area undisturbed as well. The contractor who was appointed to relocate the pipelines decided on a risk-driven verification of the suitablility of this option and a GeoQ approach was applied.

By starting with the first GeoQ step of gathering information, this project can be considered as some sort of greyfield project, because the location of the motorway and pipelines are more or less fixed by all kinds of federal and local regulations. In this particular project, interaction with the environment meant no disturbance of the protected nature reserve area. The existing ground data were rather limited in this early phase, prior to construction of the motorway. It consisted of only two borings at a considerable distance from the proposed river crossing, as well as a few cone penetration tests.

For this rather small project the second and third GeoQ step of risk identification and classification were performed by using the joint expertise and experience of the client, the engineer and the contractor. The presence of gravel was considered as the main ground-related risk for this particular directional boring. For those readers interested, Box 9.2 presents some background information about what may be indicated as the gravel nightmare for horizontal directional boring.

A careful analysis of the limited existing ground data, from the gravel risk perspective, showed that the quantity and quality of the existing ground data were insufficient in order to assess whether the gravel risk was acceptable or not for the contractor. The available borings were carried out to only 6 m below the surface, while the horizontal boring would be drilled a lot deeper, in order to be able to cross the river. Clearly, the borings were not deep enough. The available cone penetration tests (CPTs) did not really contribute either, as these show refusals on gravel layers, of which the thickness remains hidden. In addition, while acknowledging the geologically determined heterogeneity of river deposits, it was considered as not acceptable to interpolate between the borings. The limited depth of the borings could even be caused by refusals on serious gravel deposits. This worst case ground conditions scenario was further supported by a boring carried out at even greater distance, where a 12 m thick gravel layer was encountered at 6 m beneath the surface. If a similar body of gravel was also present at the location of the pipeline junction, then only a very expensive boring technique could be used. This would involve so-called obliquely boring through the gravel layer using an auger at the start and finish point of the boring section. Major extra costs would be the result of this situation.

Box 9.2   The gravel nightmare for horizontal directional boring

To install pipelines using horizontal directional boring, a relatively small hole is first bored along the required route. The diameter of the bored passage is then enlarged using a reamer, so that the pipeline can be pulled though. To ensure that the borehole remains stable, a boring fluid is introduced under pressure into the passage. This works well in the case of sand and clay. If gravel is present along the pipeline route, however, the boring fluid may flow away between the gravel particles and provide insufficient counter pressure. This can lead to collapse of the borehole, particularly where the horizontal sections of the bored route are located. As a consequence, the drilling equipment gets stuck and is very difficult, if not impossible, to retrieve. Obviously, occurrence of this kind of risk creates a maximum pressure on the budget, planning and quality of the project. This blocking of a horizontal directional boring, because of unfavourable ground conditions, occurs at least a number of times each year in Dutch practice. It is a nightmare to all parties involved and often results in major disputes, claims and eventually legal affairs.

Therefore, as a risk remediation measure, step four of the GeoQ process, an additional ground investigation was defined. To obtain more certainty to decide upon the very expensive horizontal boring solution, more information was needed about the ground conditions at the location of the river crossing. Balancing the risk profile of the project with the additional site investigation costs resulted in the execution of two deep conventional vertical borings, with sampling, on both banks of the river.

As part of the risk evaluation, step five of the GeoQ process, these borings were carefully interpreted. The findings from these borings were unexpected but highly desired. Only a thin gravel layer and a thick hard sand layer were demonstrated to be present by the two additional borings. These layers presented no problems when carrying out the directional boring, according to the expert opinions of the ground and drilling professionals involved. The favourable ground conditions enabled the pipeline to be installed by the proposed directional boring technique without the need for very expensive gravel risk reducing measures. These findings were reported, as the GeoQ mobilization step of ground risk data, ready for use in the following projects phases, by those parties involved.

This case study of a rather small project demonstrates a quick and easy application of the GeoQ process, without the aid of rather sophisticated tools, such as the EBR for risk identification and classification. In this case GeoQ proved to be just a structured and risk-driven process. It did not involve any extra costs, compared with the conventional activities in the pre-design phase. However, a major risk became foreseen. The necessary risk remediating measure, by a very focused and limited ground investigation, was carried out. As a result the major gravel risk was reduced to an acceptable level. Therefore, the horizontal directional drilling option could proceed in the next project phase, with an acceptable risk profile for the parties involved.

Summary

This chapter demonstrated the application of the GeoQ process in the pre-design phase of construction. It focused on the GeoQ steps of risk identification, risk classification and risk remediation. Three approaches to support ground risk management during pre-design were introduced and discussed. First, the team-based approach for risk identification and classification can be applied highly efficiently by ICT supported facilities, such as the Electronic Board Room. Second, the relationship between site investigation costs and risk has been demonstrated. Only by subjective judgement are we able to balance the project risk profile with the optimum scope of ground investigations. However, major cost-saving opportunities arise if we add risk as a third dimension to the conventional dimensions of guidelines and expertise for the definition of ground investigations. An adequate site investigation appeared to be a viable part of risk remediation and, based on its results, we should be able to select the most suitable pre-design solution.

The third ground risk management approach presented a number of considerations for a risk-driven ground investigation. It described how to balance ground investigations with risk, by defining the type, quantity and quality in a risk-driven way, while fitting the project's risk profile. Two case studies, a large tunnel project and a much smaller horizontal directional boring project demonstrated the application and value of the GeoQ approach in practice.

10 GeoQ in the design phase

Introduction

For any construction, an appropriate design serves as a foundation of success. In other words, design will make or, on the contrary, literally break the project. The previous chapter covered the project phase of pre-design and is followed here by design, either arranged by the client or performed by a contractor in a design and construct project. Figure 10.1 presents the design phase as the third step in the GeoQ process.

The project's pre-design will be worked out in detail during the design stage. If it has been decided for a bored tunnel at the end of pre-design phase, then a *bored* tunnel design will have to be worked out in detail. The GeoQ process aims effectively to support the risk-driven design activities, from a ground conditions perspective and within the project specifications and the risk tolerance of the involved parties. By the end of the design phase, the indicated bored tunnel should be ready for construction with a known and

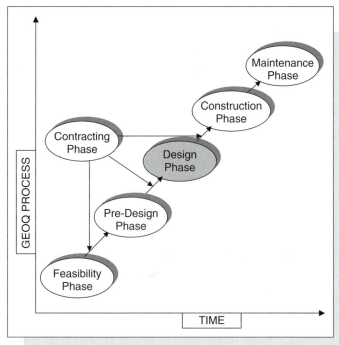

Figure 10.1: The design phase within the six GeoQ phases.

accepted risk profile. If the client is involved in design, the contracting phase will be next, in which the client brings the project to the market for tendering. If the contractor has already got the project, in the case of design and construct, this means that the design phase will be followed by the construction phase.

For reasons of simplicity, this book considers detailed design for construction as being part of the design phase, while recognizing the separation of these phases in many (large) construction projects. However, this approach does not affect the application of the GeoQ process. In *both* phases, design and detailed design, the entire GeoQ cycle should be executed, to raise maximum effectiveness.

Two methods to support ground risk management during design will be introduced and thoroughly discussed. These methods are, in particular, related to the fourth GeoQ process step of *risk remediation*, which is now of paramount importance. Any (detailed) design costs a lot of effort and resources and major design changes are occasionally (very) expensive, time consuming and frustrating to the professionals involved.

The first method identifies and selects the most appropriate remediation of foreseeable ground-related risk. General options are the reduction of the risk *cause*, the reduction of the risk *effects*, or a combination of both. After an introduction of risk cause and effect remediation, these two approaches are explored separately in more detail. The role of probabilistic methods will be highlighted for risk cause reduction, while the observational method and fall-back scenarios prove to be useful for risk effect remediation. Indeed, the concept of scenario analysis is revisited here as a risk remediation tool and in close cooperation with the observational method. The latter monitors the ground during construction for checking whether the anticipated behaviour, as used for the design, indeed occurs or not. If not, the fall-back scenarios have to be used.

A risk-driven definition and application of ground investigations is the second method of ground-risk remediation during design. Detailed and advanced ground data are often highly desired, to decide on the most suitable risk remediation measures, while this information is obviously also required for normal design activities, according to conventional design standards and specifications. A six-step model will be introduced, which breaks object-related risks down to ground parameter level. This method serves as a basis for truly risk-driven ground investigations in the design phase.

The order of application of these two ground-risk remediation methods is highly project-specific. Some projects may first need an additional ground investigation, to decide on the type of risk remediation. Other projects will explore the risk remediation options and then perform the required ground investigation, to verify the risk remediation options.

Two case studies are presented, in which the GeoQ process played a dominant role and, as usual, the chapter ends with a summary.

Ground risk management during design

Risk remediation: cause or effect approach?

Risk remediation is step four in the GeoQ process and should be preceded by the GeoQ steps of *gathering information, risk identification* and *risk classification*. After these three steps, the most serious ground-related project risks should explicitly be known. Next we have to deal with these risks. Chapter 7 introduced five risk response strategies, in order to remediate risk in some way or another: risk retention, risk reduction, risk transfer, risk avoidance and risk ignorance. Which of these specifically apply to the design phase?

This section explores the risk response strategy of *risk reduction*, which proves to be particularly useful during design when compared with its alternatives. Risk retention, by just accepting the risk and doing nothing, will be unacceptable for any serious risks. Transfer of serious risk to another party is often not possible, particularly when ground-related risk is an integral part of a design and construct contract. Only if a risk is out-of-control of the designer, risk transfer is possibly a serious option by following the golden rule of risk management, about who should be responsible for which risk. Avoidance of risk is often worthwhile considering during design, as an alternative to risk reduction. For instance, the settlement risk of a shallow foundation can be avoided by a piled foundation towards deeper and stronger strata. Obviously, the last risk response strategy of risk ignorance should not be taken too seriously anymore.

The concept of *risk reduction* demands a more detailed exploration. This approach has in fact three dimensions:

1   regarding the risk probability or likelihood of occurrence: risk cause reduction

2   regarding the risk effect or consequences: risk effect reduction

3   regarding both risk probability and effect: combined risk cause and effect reduction.

Figure 10.2 presents these three types of risk remediation within the earlier presented risk matrix.

Morgenstern (2000b) introduces the Consequential Risk Analyses (CRA) as, in his words, 'attempts to prevent a particular outcome from occurring or to mitigate the impact of the outcome'. This phrase indicates a cause and effect approach for risk reduction and Morgenstern advocates combining the systematic application of qualitative risk analysis tools, like Potential Problem Analysis (PPA) and Failure Mode and Effect Analysis (FMEA), together with the observation of ground behaviour during construction. How does this approach work in practice?

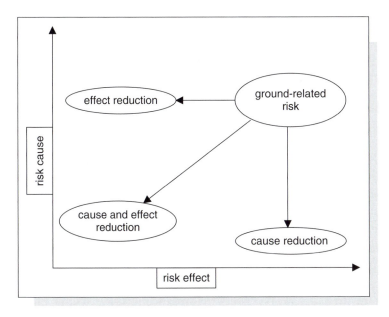

Figure 10.2: Ground risk remediation by cause and effect reduction.

From time to time, the risk of *flooding* is a serious threat to many people all over the world living in relatively low-lying areas. This risk of *flooding* provides an illustrative example to the possibilities for risk cause and risk effect remediation.

For reasons of simplicity, let us limit the risk cause to the failure of a water retaining structure, such as a dike. The strength of these structures is highly dependent upon the underlying ground conditions. An aspect such as *piping* is a failure mechanism, in which groundwater flows create some sort of pipes or holes in sand layers underlying the dike. These may seriously undermine the dike stability and, in situations with extremely high water levels and the resulting extreme loading conditions, such a dike may fail.

The *cause reduction* of the risk of flooding involves the reduction of the cause part of the flood risk. Many Dutch dikes are designed to an extremely high water level of once in the 10 000 years. Consequently, the dike should be strong enough to resist this extreme high water and its associated loads. This evidently results in rather heavy dikes that, in fact, turn out to be over-dimensioned for 9999 years during the 10 000 year's period.

Alternatively, risk *effect reduction* involves measures to reduce the effects of dike failure. Therefore, in the early days, the Dutch living in high-risk and low-lying areas furnished the ground floors and walls in their houses with tiles, to about 1 m height. If high water conditions were expected, typically to be judged by the weather conditions in those times, then they moved their precious furniture to the first floor. In the case of a flood, which was quite normal, they moved themselves to the first floor as well. As a result of the tiled floors and walls, the water-damage was limited and needed mainly cleaning. Obviously, if there was a serious flood with water levels considerable higher than 1 m, these brave people got into trouble. Large-scale evacuation programmes are more modern equivalents to the ancient risk effect measures against flooding.

The combination of the cause and effect reduction implies the reduction of both risk dimensions. In this example it means both strengthening the dikes and preparing large-scale evacuation programmes. Box 10.1 presents the Dutch tendency to pay more attention to risk effect reduction with regard to the risk of flooding.

Box 10.1   Revival of risk effect reduction

Recently, in The Netherlands, the aspect of flood risk effect reduction gets more attention, in particular from the Dutch ministry that is responsible for flood protection. This revival is caused by the increasing lack of space in The Netherlands. Many local governments want to develop domestic areas in the abundant low-lying parts, close to the main rivers and coast. However, insurers are not ready to insure the risk of flooding, in spite of the fact that many people want to live in these areas. The costs for upgrading dikes to the modern safety standards are enormous, apart from a lot of resistance by all kind of environmental pressure groups, who want to protect the typical Dutch landscape. This situation triggers the development of innovative construction methods that serve as a flood risk effect remediation. Examples are houses on piles above ground level, as well as the design of *floating* houses. Indeed, this type of construction has already been common practice for a long time in other parts of the world.

Both risk cause and effect reduction need due consideration in the design phase, even while risk effect reduction is typically implemented during the construction and maintenance phase. The choice to remediate *after* design, however, should be made already *during* design. This results, for instance, in a number of fall-back scenarios, to be agreed upon during design and to be applied during construction, when required.

If cause reduction, effect reduction and even their combination is considered as not appropriate for a particular risk, we may consider the remaining response strategy of *risk avoidance*. I recall the example of the flood risk and the dike for a new domestic housing project in a low-lying area. If both dike strengthening and specific construction measures against flooding are likely to become over-expensive, then relocation of the project to a less flood-sensitive area would be required. This need for relocation has preferably already surfaced by the risk analyses during the feasibility or at least pre-design phases. While forced to take such a relocation measure during the design phase, the risk management process typically failed in the previous project stages. However, it is even better to become aware of it during design, rather than just continuing the project in unawareness. This situation would result in an unacceptable flood risk during the operation phase of the domestic housing project, when the people actually live there.

Finally, also the type of construction contract often influences the selection of risk reducing options. In a design and construct project the contractor is responsible for both design and construction. Therefore, it is up to the contractor to decide upon ground risk remediation during design or construction and the contractor will benefit from it. In a traditional type of contract, with a clear distinction between design and construction, it is much more difficult to anticipate risk reduction during construction, simply because it is the responsibility of another party.

Reducing the risk cause: ground parameters and the probabilistic approach

If we choose to remediate a ground risk by its *cause* reduction, we have to look the inherent ground uncertainties straight in the eyes. We have to deal with the fuzzy, random and incomplete ground character. In this respect, the earlier introduced methods of Fault Tree Analysis (FTA), Failure Mode and Effect Analysis (FMEA), and Failure Mode, Effect and Criticality Analysis (FMECA) help to break the risk causes down to the level of individual ground parameters. However, these parameters, such as the undrained shear strength, rock discontinuity shear strength, soil permeability, *vary* inherently, as many readers may have experienced from their ground investigation, engineering and construction involvement. Which values of these parameters should be applied in the design in order to arrive at the desired risk cause reduction, without expensive over-dimensioning? A *probabilistic* approach may help to deal effectively with this inherent parameter variation.

What is a probabilistic approach? Let us first have a small look backwards. Since the development of the ground engineering disciplines, somewhere in the 1920s, the so-called *deterministic* approach has been widely adopted in ground engineering. It assumes an explicit and unique answer to a geotechnical problem, either by the application of theoretically sound relationships, or by empirical or semi-empirical methods (Bell, 1987). However, as we already had to accept earlier, explicit and unique answers are non-existent when dealing with ground. Ground-related answers depend on a number of uncertain factors and are subject to a certain *probability*. A common definition of probability is a numerical measure of the likelihood that an event will occur. This probability measure is often expressed in a percentage, like 20 per cent or 80 per cent. The concept of probability may help us in measuring, expressing and analysing the uncertainties of future events, as expressed by Anderson et al. (1999). If we were really sure about a ground-related answer, then the probability of that answer would be 100 per cent. In reality, we have to live with answers typically far below this 100 per cent probability. As stated by Bell (1987), a probabilistic approach adopts the concept that several outcomes are possible. Therefore, a *probabilistic* approach can be of great help to deal explicitly with uncertainty in design. Authors, such as Lacasse and Nadim (1998) and Ho et al. (2000), who also advocate a wider application of this approach, confirm this statement. Typical ground-related aspects covered by probabilistic methods, which became available in the 1990s, are groundwater flow, settlements, slope stability assessments and the consideration of liquefaction potential (Hicks and Samy, 2002).

However, because of its new and rather complicated nature, probabilistic methods still remain in the neighbourhood of highly educated and experienced specialists. More easy to handle software will certainly contribute to a much wider application and, in the near future, I foresee a sharp increase in the application of probabilistic approaches within the GeoQ ground risk management process.

Obviously, there are some reservations to be made. The large amount of reliable ground data, as required for a lot of probabilistic analyses, is often still lacking, in particular in the smaller construction projects. Furthermore, as stated earlier, we should not over-rely on this apparent objective approach that is inherently driven by subjective expert judgement. However, we should not miss the opportunities of probabilistic methods, as these may at least provide some minimum but explicit level of certainty about the most relevant ground parameters for the project. Often, these parameters have proven to be the drivers of many of the identified and classified as serious ground-related project risks.

Recent evolutions in the approach to the *safety factor* demonstrate the irreversible advance of probabilistic methods in the ground-related disciplines, as presented in Box 10.2.

First the good news about this evolution of the safety factor. The total ground uncertainty is split up and allocated to the relevant ground parameters and the loads to be used in the design calculations. Also geological heterogeneity of the ground explicitly gets attention. Now the bad news; this approach appears to cause a lot of confusion and controversy, because of the lack of knowledge and experience about the way to arrive at quantified values for these partial factors of safety (Hicks and Samy, 2002). According to Hannink et al. (2004), research within the Dutch geotechnical community resulted in similar findings. Not being fully familiar with the ins and outs of recent geotechnical guidelines resulted in project inefficiencies, conflicting opinions between the different ground specialists and unnecessary conservative design with hidden safeties. Obviously, unnecessary costs are involved with over-dimensioned design. The probabilistic approach and its methods are apparently not yet fully adopted and used within the geotechnical community.

Box 10.2   The evolution of the safety factor

The *conventional* factor of safety is used in the well-known *deterministic* type of geotechnical design calculations. In theory, a safety factor above 1 will be sufficient. In geotechnical practice, however, safety factors vary normally between approximately 1.5 and 4.0. These values combine all uncertainties about ground conditions, loading conditions and calculation models (Bell, 1987).

A piled foundation design is considered safe and acceptable, if the ultimate bearing capacity of the piles is at least two times the actual maximum load on the piles, which results in a safety factor of 2. Since the 1990s, the conventional and deterministic safety factor concept became step-by-step replaced by the concept of *characteristic values* for the geotechnical design process. In fact, it is the announcement of the (semi-) *probabilistic* design in ground-related engineering. The Eurocode 7 of the European Union has adopted this principle of characteristic design values, which are based on 95 per cent reliability. In other words, the probability that the actual properties deviate adversely from the design values is 5 per cent (ENV, 1994). Within the concept of characteristic values, each ground parameter to be used in a certain calculation model will have its own so-called *partial factor* of safety.

While considering the *characteristic values* of ground properties to be selected for an appropriate design, we should realize that we are dealing with *speculative risks*. It means that a ground-related risk may increase or decrease with a similar or opposite variation of the main ground properties that drive the risk. For instance, the risk of slope instability will be caused, among other factors, by the undrained shear strength of the slope material and underlying ground layers. If higher values of undrained shear strength can be used in slope stability calculations, the risk of slope instability will decrease, while applying in this simplified example the *ceteris paribus* principle, that all other factors remain constant.

A well-known question arises again: does more or other ground investigation data allow us to use more favourable ground parameters in our calculations, which will result in a reduced risk? Hannink et al. (2004) dare to give an answer: more ground data involve often a major potential for design optimizations. By catching the opportunities, these improvements reflect the positive side of risk management. The same design reliability may be reached at lower net cost, which is the reduced project cost including the additional cost for the extra ground investigation. The second case study of this chapter is dedicated to a probabilistic design approach, which has been used to select more favourable ground parameters with still an acceptable reliability. This approach resulted in attractive cost savings within the project.

The decision about what type of design to apply, deterministic or probabilistic, has meanwhile one well-known answer: *it depends*. The ground-related risks, as retrieved from the previous GeoQ steps of information gathering, risk identification and risk classification, need to be carefully analysed, in order to select the most appropriate design method with regard to the risk cause remediation. In many projects, probabilistic design certainly will add value, because it provides insight into the main ground-related drivers of a particular risk. This information may help in the process of deciding upon the main risk remediation options of risk cause reduction (during design), risk effect reduction (during construction) or risk avoidance (during design or construction). However, we should realize that the latter option will induce new ground-related risks, as ground uncertainty is always there.

I consider it beyond the scope of this book to explore further in detail the many available probabilistic calculation methods, like Monte Carlo analysis, because there are many and detailed books available about this topic. For instance, Bell (1987) presents an accessible introduction into statistics and probabilistics in geotechnical engineering. In addition, Anderson et al. (1999) provide a complete and thorough description of the discipline of statistics and probability, which extends to business and economics.

In conclusion, it is up to ourselves and the probabilistic experts we may consult, to decide whether probabilistic approaches add value to the ground-risk management activities during design of construction projects. This responsibility requires some basic understanding of the science and art of probabilistics. At least, we should be able to communicate effectively with the experts on probabilistics, to check and balance *their* results in the decision process for *our* most appropriate risk remediation measures. Let us now have a look at the other side of risk remediation: reducing the *risk effect*.

Reducing the risk effect: the observational method and fall-back scenarios

Let the ground-related risk happen, but in a controlled way. This, in fact, is the core of the second risk remediation option: reducing the *risk effect*. Box 10.3 presents an example.

In essence, Box 10.3 illustrates the application of the *observational method*, as widely introduced by Peck (1969). By combining fall-back scenarios and monitoring during construction, it is possible to challenge the foreseen limits of ground behaviour within an acceptable risk profile. However, while applied during construction, the *decision* to use the observational method needs to be taken early in the design phase. While designing, less conservative ground parameters have to be applied, for instance for slope stability and settlement calculations. This is the ultimate reason for discussing risk effect remediation during construction in this chapter about the design phase. The example of Box 10.3 aims to act as an appetiser for a case study in Chapter 12 about the construction phase, where the observational method in combination with fall-back scenarios has been successfully applied.

Box 10.3   Let the ground risk occur, but in a controlled way

A road construction project provides an example of a controlled ground risk that may occur without problematic effects. Two risks of concern are the slope instability of an embankment that is constructed on rather soft soil and the embankment settlements. During construction, the height of the embankment is restricted by the undrained shear strength and the consolidation behaviour of the underlying soft soil layers. The load of the embankment causes groundwater overpressures in these layers, which will temporarily reduce their effective stresses and strength.

   By the conventional way the contractor could only start with a 2 m high embankment, to prevent embankment instability. After a certain period of time, often in terms of several months in which the groundwater overpressures in the underlying ground dissipate to lower levels, the next layer of 2 m can proceed.

   However, the faster the embankment can be constructed, the earlier the embankment settlements will have occurred and the more *certainty* the contractor will generate about meeting his residual settlement requirement. This demands balancing the risk of temporarily embankment instability, during construction, against the risk of not meeting the settlement requirements at the end of the project. By reducing the *risk effect* of the embankment instability, it may be allowed to hasten its construction, in favour of reducing the settlement risk. One or a few *fall-back scenarios* for the embankment instability are required and include the methods, time and resources to be reserved for repair activities. These fall-back scenarios provide a set of measures to be executed in case the risk of embankment slope instability effectuates. By *monitoring* changes in groundwater pressures in the underlying ground layers, before, during and after the installation of the embankment fill, valuable information about the ground response to the embankment will be gained, which might even allow the embankment construction to be further speeded up. In addition, the settlement behaviour can be monitored by measuring the vertical deformation at and in the vicinity of the embankment. Any effects of the embankment construction process on the settlements become visible and by back-analysis the expected residual settlement predictions can be refined. In this way, besides controlling the embankment instability the effect of the settlement risk is also being controlled.

For the less serious ground-related risks, in particular, reducing their effects instead of their causes can be cost-effective. Hidden opportunities of *better* ground behaviour than expected, which can only be revealed during construction, are used by this approach. Contrary to the concept of risk cause reduction, the majority of risk remediation costs are only made when the risk indeed occurs, obviously apart from costs made for the preparation of fall-back scenarios and the execution of additional monitoring. Preparedness for surprise, change and adaptation during construction are, however, definite prerequisites for the application of risk effect reduction as a remediation measure. This project flexibility during construction needs to be explicitly anticipated and built-in by suitable design.

Chapter 8 discussed scenario analysis as a risk identification tool in the feasibility phase. Here we revisit the concept, to apply it as a fall-back scenario for supporting risk effect reduction. The types of *range of futures* and *alternate futures* scenarios can both be applied as fall-back scenarios. The first provides a range of possible outcomes. An example is a range of increasing damage to adjacent buildings, from cracks to even collapse, caused by adverse horizontal deformations due to construction pit excavation. The risk effect reducing measures will depend on the degree of horizontal deformation. They can vary from just repairing the minor crack to the installation of (additional) horizontal struts in the construction pit, when the horizontal deformations grow larger, to ground improvement by grout injection techniques outside the construction pit, after stopping excavations in case of real unexpectedly large deformations. In this case, the risk of unacceptable horizontal deformations should already have been considered as a foreseeable risk during design. Consequently, an unnecessarily heavy and expensive sheet pile wall can be avoided. Monitoring will be required during construction in order to measure the horizontal deformations during the excavation of the construction pit. All these risk effect reducing measures demand explicit consideration in the design phase of the construction and, obviously, the cost savings of the lighter sheet pile wall have to be judged in view of the probability and costs associated with the range of possible horizontal deformations.

The second scenario type of alternate futures provides a number of *discrete* outcomes. Let us consider dewatering of a project site to allow for construction in dry conditions. In this example, dewatering may reduce the groundwater level in the project area to a considerable distance, which causes damage to crops of a number of farmers. A discrete fall-back scenario is to use the pumped water for irrigating the crops. Another scenario is to compensate the farmers financially for having no crops during one season.

Obviously, each fall-back scenario has its costs and possible other unfavourable side effects, such as demanding a monitoring programme and delay when the work has to be stopped in case the fall-back scenarios become reality. Therefore, reducing the risk effect should be carefully compared with the alternative of reducing the risk cause. This judgement needs seriously to take place during design. As many readers probably recognize and even may have experienced, major changes during construction are normally highly unwanted, because of their often enormous costs and planning complications. This brings us to the next section, because detailed and advanced ground investigation proves to be another viable remediation measure in order to minimize the risk of major changes during construction, caused by inappropriate design.

Detailed and advanced ground investigations

The previous chapter presented the main objective of the preliminary or, rather *general*, ground investigation in the *pre-design* phase: to arrive at a conceptual model of the ground. This model reveals the main ground types, together with the main geotechnical, geohydrological and geoenvironmental mass and material properties, within the risk profile as available during pre-design. In the next phase of *design*, the most important ground-related risks for the remaining project phases need to be disclosed, including their main risk-triggering ground parameters. Usually, major design decisions in relation to ground risk remediation measures have to be taken and the effects of these decisions echo through all the remaining project phases. Additional ground information is needed to complete the risk-driven design in detail, with ample attention to ground-related risk management, either by risk cause or by the risk effect remediation. This calls for *detailed* ground data to be retrieved by sensitive and *advanced* ground investigation techniques. The results of these detailed ground investigations colour the white spots, the unknown areas of ground properties and behaviour, which doubtlessly surface at a certain moment. Therefore, the second method for ground risk remediation concerns *how* to define a balanced and risk-driven ground investigation programme during design. Box 10.4 presents six steps to arrive at such a programme, by a breakdown of risks to ground parameter level. (It presents an example in parentheses.)

These six steps in Box 10.4 provide a receipt that more or less automatically results in a risk-driven ground investigation. An additional risk management loop is, in fact, made for each ground-related object of the project. Step three may require a further cause and effect breakdown of risk, possibly by using the FTA, FMEA or FMECA risk analysis tools. By this in-depth risk analysis of all ground-affected constructions and the resulting ground investigation, a number of hidden and critical risks, as well as opportunities, are likely to surface. Their remediation measures can be incorporated in either design or construction.

Box 10.4   Six steps to arrive at a risk-driven ground investigation

1   Determine the ground-related *constructions* of the project (ground embankment for a road that serves as connection with a bridge on piles)

2   Determine the main geotechnical, geohydrological and geoenvironmental *mechanisms* that affect the fit-for-purpose of these constructions (unacceptable large settlements of the embankment)

3   Determine the *risks*, assessed by probabilities of occurrence and effects, if the identified geotechnical, geohydrological and geoenvironmental mechanisms act adversely (the probability and effects of unacceptable large settlements)

4   While considering the risks of step 3, determine the most appropriate *design techniques* for the geotechnical, geohydrological and geoenvironmental mechanisms (finite element settlement calculation)

5   While considering the selected design techniques of step 4, determine the most critical *ground parameters* (soil stiffness for the finite element settlement calculation)

6   While considering the ground parameters of step 5 and the anticipated *geological heterogeneity* on the site, determine the type, quantity and quality of the ground investigation (10 borings with undisturbed piston sampling and 30 constant rate of strain deformation tests in the laboratory).

Probabilistic approaches can be added to the model of Box 10.4. For instance, Calle (2002) demonstrates how a probabilistic approach can be used to decide on the optimum *quantity* of ground investigations. By his approach, costs of ground investigations are compared with the probability of achieving the objectives of the ground investigation. An example is the detection of sandy channel-deposits by Cone Penetration Testing (CPT). These deposits are highly variable, as a result of their alluvial or river-deposited geological history. We recall the risk of *piping* below dikes, which is highly dependent on the presence of these underlying sand layers. Calle (2002) balances the probability of locating a channel deposit with the number of CPTs. More certainty about the channel deposits will demand more CPTs. The cost of remediating the piping risk by a dike improvement is compared with the cost for the site investigation and the probability that sandy deposits are located. This approach allows restriction of the piping risk remediating measures to those areas were the channels are located with an acceptable reliability. Consequently, dike strengthening can be avoided in those sections were the channel deposits are not located by the CPTs, again with a certain agreed reliability. The

Table 10.1: Increased safety factors by a set of local ground parameters (Greeuw and Van, 2003)

Ground types	Safety factor against slope instability	
	By Dutch design code NEN 6740	By a local data set of 12 triaxial tests
Clay: Wad-deposit	0.95	2.15
Dike material: Schieland deposit	0.71	1.74
Clay: Duinkerke deposit	0.95	1.82

Note: Stability factors are calculated by the method of Bishop in MStab software

appropriate number of CPTs is, in fact, the key to this approach, which may result in significant cost reductions in dike reinforcement programmes.

Detailed ground investigations can also be used to provide databases with local and project-specific ground properties. These properties may allow the application of more favourable values of ground parameters in design calculations, when compared to parameters as recommended by guidelines. Greeuw and Van (2003) demonstrate the value of a rather limited local data set. A relatively small set of 12 triaxial tests resulted in a safe application of much more favourable ground parameters when compared with the necessarily conservative characteristic values of the same parameters in the design codes. The values from the design codes were considered as safe, but as (too) conservative as well, which would result in an over-dimensioning of a slope design. Table 10.1 presents the increase of the safety factor, for three soil types, by using the results of the triaxial tests for a slope stability design.

As demonstrated in Table 10.1, slope stability calculations with ground parameters derived from the design code resulted in safety factors below 1, which indicate an unstable slope. By the application of the local data set, the safety factor increased to values around 2, reflecting an acceptable safety against slope instability. This example resulted in a much more cost-effective slope design and demonstrates the potential benefits of *locally* better ground conditions, compared with the necessarily conservative assumptions based on *regional* data sets and *national* design codes. The application of *detailed* ground data during design is therefore not only favourable to risk reduction, it may pay off with respect to value engineering as well. More favourable geotechnical design parameters may be allowed to be used, while retaining the same pre-set reliability and safety standards. This implies reaching a pre-set design reliability at lower net cost, which is corrected for the extra ground investigation cost.

Apart from more detailed ground data, the *quantity* aspect, the application of more advanced site investigation tools may also be attractive, the *quality* aspect. Box 10.5 presents some concise examples, drawn from my own experience, of using a few relatively advanced ground investigation techniques in projects in different parts of the world.

Box 10.5   Detailed and advanced ground investigations for risk remediation

In Singapore, the behaviour of *weathered rock* and *residual soils* played a dominant role in the foundation design for a chemical plant with near-shore facilities. Advanced triple tube core drilling has been specified and performed, because the properties of these fuzzy materials contributed largely to the risk of unacceptable differential settlements.

In West Africa, the foundations for Liquid Natural Gas (LNG) storage tanks needed detailed attention. Also for this project the risk of unacceptable settlements demanded a detailed ground investigation. Very heterogeneous soft soil deposits in a delta area, with a thickness over 40 m, were investigated by electrical Piezo Cone Penetration Test (PCPT) techniques. The PCPT measures not only the soil's strength properties, but also pore water pressures. By so-called *dissipation tests*, the reduction of groundwater overpressures over time has been measured. This detailed geotechnical and geohydrological information added value to the design of appropriate settlement-reducing measures for the LNG tanks.

The extension of a coal terminal in Indonesia required a detailed assessment of the soil liquefaction risk, because of its location in an earthquake-prone area. During a detailed site investigation, conventional Standard Penetration Tests (SPTs) have been complemented by PCPTs. By comparing the SPT and PCPT results, the contractor was able to perform a more detailed and reliable assessment of the site's liquefaction potential. The findings resulted in an appropriate design by piled foundations of the industrial facilities, as a risk cause remediation measure.

I immediately admit that both the site investigation tools and their application, as presented in Box 10.5, are probably not brand-new to many readers. However, their application occasionally still requires considerable debate with clients. It is often rather difficult to communicate explicitly their added value to decision-makers who are less familiar with ground conditions. The risk-driven motivation to apply these techniques may release them from their reputation as 'too expensive for my project'.

Also in the field of geoenvironmental engineering, sophisticated tools, like CPT technology combined with soil resistivity measurements and groundwater sampling, proved to add value for geoenvironmental risk remediation. Chapter 13 presents a case study about this topic. Evidently, risk-driven and sophisticated laboratory investigations, such as triaxial tests with unloading and reloading cycles and model testing in geocentrifuges, may significantly contribute to risk reduction during design for particular projects all over the world.

The indicated advanced ground site investigation and laboratory techniques are just a fraction of the many globally available tools and techniques. I consider it beyond the scope of this book to provide comprehensive lists of all technologies available, together with their advantages and disadvantages. Specialist advice should be sought in this respect in order to arrive at the most cost-effective ground investigation programme during the design phase. Besides satisfying the conventional demands for safe and reliable design, detailed ground investigations with advanced tools also provide the essential information to remediate ground-related risk, either already in the design phase or in the next construction phase. This latter function of detailed ground investigations seems not yet always explicitly realized by the professionals involved. In view of the presented six steps approach for defining risk-driven ground investigations, advanced ground investigation tools offer ample opportunities for both risk remediation and value engineering within the GeoQ risk management process. The next section of this chapter presents two case studies, in order to support this statement.

Case studies

Several concise cases, from all over the world, have previously been presented. This section presents two Dutch case studies in some more detail. They concern the *design* of (parts of) a project. The purpose of the cases is to demonstrate the possible application of the GeoQ process in the design phase of construction projects.

Liquefaction risk control below a railway

The area of one of the world's largest ports, Rotterdam, provides locally probably the worst soft soil conditions in The Netherlands. In this area, from the city of Rotterdam towards the Dutch political capital of The Hague, a light-rail link will be constructed. The project will be realized in a bored tunnel between the Rotterdam Central Train Station and the northern part of the city of Rotterdam. The tunnel is going to cross the existing Rotterdam–Gouda (well-known for its cheese) railway at 14 m depth. This particular part of the project is the topic of this case study.

The project site can be considered as a typical greyfield, by applying the first GeoQ step of *gathering information*. The location of the tunnel is fixed and characterized by an existing and intensively used railway above the tunnel. According to a conventional method in the past, the existing Rotterdam–Gouda railway appeared to be constructed by the so-called continual pressure method: sand has been deposited at the proposed alignment until it simply no longer settled. This method resulted in a water-saturated foundation layer of rather loosely packed sand. Existing ground data revealed a 12 m thick, loose man-made sand bed, which pushed away the originally very soft peat layers. Figure 10.3 shows a cross-section with the soil profile and a typical Cone Penetration Test (CPT) result.

Based on this existing information, the second and third GeoQ steps of ground *risk identification* and *risk classification* were performed. While existing ground data, by the CPT-results, proved to be of great help, the available information appeared insufficient for acceptable risk remediation.

Geotechnical experts of the client's design team identified a serious risk. During the boring operation, the tunnel-boring machine would generate vibrations, which might liquefy the loosely packed

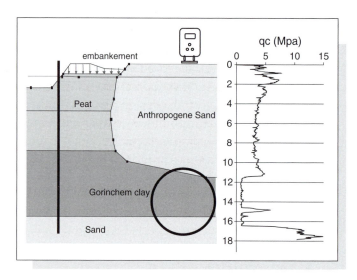

Figure 10.3: A typical cross-section of a railway on soft soil.

and saturated sand underneath the existing railway. The occurrence of unacceptable degrees of deformation of the existing and busy railway line raised concern. This liquefaction risk was classified as highly serious, as it could cause substantial damage along the existing railway. The causation of the liquefaction risk provides a relationship between the density of the sand and the sensitivity to liquefaction. Generally, the liquefaction potential of a site increases with a decreasing density of the sand deposits underlying the site.

During the fourth GeoQ step of *risk remediation* the project team had to decide between two main risk response strategies: risk avoidance or risk reduction. Avoidance of the entire risk by another tunnel construction method was considered as unacceptable. In the densely populated area a bored tunnel has a lot of advantages, including minimum disturbance of the city life in the neighbourhood of the tunnel. Regarding the remaining option of risk reduction, possibilities for both cause and effect reduction needed appraisal and judgement. Reducing the risk cause, the vibrations generated by the tunnel-boring machine, was considered as not feasible and thus the risk response by effect reduction remained. A temporary steel bridge to support the existing railway was proposed as a promising solution. The application of soil improvement techniques alongside the existing railway line could act as a fall-back scenario, in case the temporary bridge could not entirely fulfil its function.

The resulting risk profile was carefully analysed by interpreting all existing ground data in the next step GeoQ step of *risk evaluation*. However, the rather loosely packed sand deposits below the existing railway link remained a source of uncertainty. Although conventional electrical CPTs are reliable for the assessment of soil conditions, the correlation of its results with the in-situ sand density is known to have quite a wide margin of uncertainty. Therefore, in close cooperation with an external geotechnical consultant, the project design team decided to define a detailed and advanced ground investigation. This was expected to pay off in terms of a more reliable liquefaction risk remediation. It was decided to perform in-situ electrical density measurements, by using small-sized conventional cone penetration test equipment. The tests were performed under an inclination of 30 to 45 degrees below the existing rail track, adjacent to the existing railway that could stay in full operation during the ground investigation activities, as shown in Figure 10.4.

With the same small-sized CPT-equipment it was possible to take a number of sand samples for calibration purposes in the laboratory. Box 10.6 dedicates a few sentences to this advanced type of ground investigation, for those readers who are interested.

Figure 10.4: In-situ and inclined electrical density testing below a railway (© with permission of GeoDelft).

Box 10.6   Advanced ground investigation by in-situ density testing

The in-situ electrical density method measures continuously the electrical conductivity of the soil and the groundwater with two specially equipped cones, a soil cone and a groundwater cone. The soil cone has two sets of electrodes for sending and receiving an electrical current. The water cone measures the conductivity of the groundwater. The cones are operated by conventional cone penetrometer equipment, which is widely used for Cone Penetration Tests (CPTs). By the same CPT equipment it is possible to retrieve samples, by several available sampling techniques, including the so-called Begemann system and MOSTAP systems. The test results are converted to in-situ density, after measuring the conductivity of the sand samples with different densities in the laboratory.

Additional laboratory testing was performed to determine the sand deformation properties under shear stress, by simulating the stress conditions during the passage of the tunnel-boring machine. The results of the additional ground investigation were beyond expectations. Even an alternative design for liquefaction risk remediation came up. It involved a combination of only soil improvement techniques, without the need for the temporary bridge. These proven techniques included increasing the height of the sand alongside the existing railway line, stiffening the soil in the vicinity with lime-cement columns, and strengthening the sand layers by using gel-injection techniques. Advanced model calculations were performed to test these alternative risk remediation measures. The tests and calculations indicated only limited deformations, which would take place rather gradually. By means of monitoring and maintenance it would be feasible to control any minimum subsidence of the existing rail line. After balancing all possible options, in close cooperation with the geotechnical consultant, the project team decided to omit the temporary bridge entirely and to apply only soil improvement measures (Korff, 2003). The evaluation of these risk reducing methods resulted in an acceptable risk profile for the parties involved. Avoidance of the temporary bridge saved a few million euros in risk control costs. In addition, the temporary bridge would have required interruption of the operation of the existing and busy railway during at least two weekends, to install and remove the bridge. This unfavourable side effect has also been eliminated. The detailed and advanced risk-driven ground characterization demonstrated, therefore, how to reduce the perceived liquefaction risk in a cost-effective way by defining of a number of proven techniques to control the risk effect. This may be considered as an example of risk-driven value engineering in the design phase. Finally, as the last GeoQ step, all ground risk information gathered had to be filed in a risk register, so that it could be used effectively by the contractor during the construction phase of the project.

Settlement risk control under pressure

Many urban areas in the world are located in deltas at or near the coast. A complex geological history with subsequent alluvial and marine deposits has often one characteristic in common: these deposits are soft and will cause serious and long-term settlement when loaded by almost any type of construction. Piled foundations are a way to deal with these settlements. However, for infrastructure projects like railways and roads, piled foundations over distances of several tens to hundreds of kilometres are very expensive, if not unfeasible.

The last case study of this chapter demonstrates the application of the GeoQ ground risk management approach for a railway project in its last part of the design phase, just before the start of construction. The geotechnical considerations are deliberately presented in some detail. The geotechnical professionals among the readers will probably recognize many of the presented issues of ground concern. I believe, however, that this case also will interest those readers who are less familiar with geotechnics and particularly read this book for learning how to deal effectively with those ground professionals within their projects. For the latter group of readers this case aims to demonstrate how an innovative geotechnical approach can be communicated, by means of risk management, towards any less geotechnically underlain project stakeholder. These people may very well act as the decision-makers of the project and have to judge and balance all these ground-related risks, as being responsible for the project's success. The following case demonstrates how rather subjective but sound engineering judgement can be made explicit and more objective, in order to arrive at a cost-effective reduction of serious ground risk.

The Betuweroute is a new double-track and electrified freight railway. The line has a length of 160 km and connects the Rotterdam harbour with Germany, as part of the European Network of Freeways. The western part of the Betuweroute passes through typical Dutch polders with very soft subsoil conditions. In the section between the cities of Sliedrecht and Gorinchem, the poorest soil conditions and tightest contractual time span met each other. This 22 km long part of the Betuweroute was contracted as a Design, Construct and Maintenance (DCM) contract to a consortium of contractors. An alliance agreement, with a joint risk management budget for the client and the contractors, was the driving force to pay rigorous attention to subsoil risk management (van Staveren, 2004).

The consortium based its initial design on a rather traditional approach of using a sand-fill embankment and so-called wick drains. This price-competitive solution resulted in the winning bid. In the final design phase, however, the contractor's consortium and the client had to take a few major decisions regarding their risk control of the ground.

With regard to the first GeoQ step of *gathering of information*, this site can be considered as typically *greyfield*, because it was located very close to an existing

Figure 10.5: Challenging working conditions during day and night (© with permission of GeoDelft).

railway and the busy motorway A15. Figure 10.5 demonstrates how parts of the project needed the motorway to be constructed. To disturb the traffic to a minimum, a lot of night shifts could reduce the traffic disturbance to a minimum.

Abundant ground data were available during the final design phase of the project. Prior to contracting, the client had already conducted a very extensive ground investigation programme, including 658 CPTs, 145 undisturbed and continuous so-called Begemann borings, 201 triaxial tests and 198 oedometer tests. One of the ground layers of concern is the so-called Gorkum clay deposit. The derived settlement properties of this deposit indicate a very slow settlement process that is expected to continue for many years. The interpretation of the available ground data in the final design phase highlighted the extremely poor subsoil conditions of the project.

During the next two GeoQ steps of *risk identification* and *risk classification*, the occurrence of unacceptable post-construction settlements, typically larger than 0.3 m, was one of the major risks that needed some sort of remediation. The causation of the settlement risk indicated, particularly, the contribution of *creep* to the final settlements as being a major uncertainty. This creep mechanism is a consolidation-independent ongoing ground deformation as a result of a load. The consortium selected the combination of both risk cause and risk effect remediation as the main risk response strategy. In the final stage of design, the cause of the settlement risk would be evaluated by a settlement prediction model. During construction, the occurring settlement would be carefully monitored and a number of fall-back scenarios to accelerate the settlements, when required, were prepared.

However, a new problem arose, the Dutch state-of-the-art model for settlement predictions was judged as inadequate by the project's team ground experts. In spite of its wide application, the so-called Koppejan model provides rather inaccurate predictions of creep settlement (Molendijk et al., 2003). The seriousness of the settlement risk and the subsoil creep as the main risk driver demanded a new and innovative model for improved assessments of creep behaviour. Based on the work of Den Haan (1994) and Yin and Graham (1999), a so-called isotache model was developed.

After the application of the risk reducing isotache creep–settlement model, the remaining risk profile was judged in view of the GeoQ step of *risk evaluation*. The expected post-construction settlements were subjected to a Monte Carlo simulation on 10 000 sets of parameters. However, the Monte Carlo analysis resulted in a probability of 70 per cent of not meeting the post-construction settlement limitation of 0.3 m. Figure 10.6 illustrates the results of the Monte Carlo analysis.

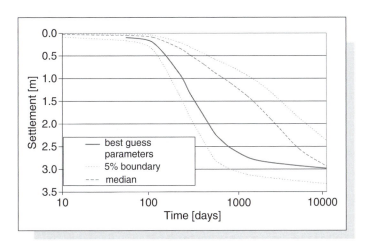

Figure 10.6: Range of calculated settlement curves by Monte Carlo analyses (© with permission of GeoDelft).

In other words, the probability of meeting the settlement requirements was only 30 per cent, based on the 10 000 executed calculations. This low probability of success was considered as not acceptable by the project's consortium. Therefore, at that moment cause reduction of the settlement risk, by the application of an innovative calculation model, failed. However, based on sound engineering judgement of the involved ground engineers and their local experience in the area, the contractor's consortium remained strongly confident of being able to meet the settlement criteria. They even kept relying on their conventional and rather cheap construction method that helped them win the contract: the combination of pre-loading by sand-fill embankment and wick drains, to accelerate the consolidation part of the settlement. In which way could this rather subjective risk perception of the geotechnical project team become rationalized to a more objective approach? How to prove their rigorous engineering judgement? It demonstrated the need for an even more reliable ground characterization, with the most reliable subsoil information possible: actual monitoring data. Therefore, the results of a recent Dutch research project, performed in the Delft Cluster Research Programme, were applied in practice. It was decided to add early available construction monitoring data in the Monte Carlo analysis, by following the procedure as described by Hölscher (2003). Here the approach of the observational method entered the project which was already in the detailed design phase. To derive the actual consolidation coefficient from field settlement data, Asaoka's method was applied, as proposed by Dykstra and Joling (2001). It appeared to work: excellent fits between the isotache creep–settlement model and the recorded settlements were found. Figure 10.7 presents the mean value and the 5 per cent and 95 per cent boundaries, indicating the 90 per cent probability interval for the settlement curves.

The range of the 90 per cent probability interval from Figure 10.7, which includes the early monitoring data, is much smaller than the similar range in Figure 10.6 without use of monitoring results. Therefore, adding monitoring data, representing the real ground behaviour in practice, into the Monte Carlo analysis decreased the uncertainty in meeting the settlement requirement dramatically. By the end of this extensive GeoQ step of

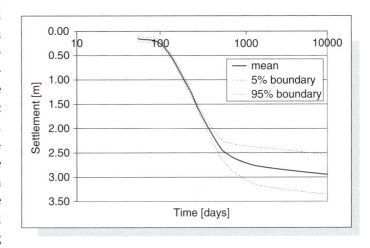

Figure 10.7: Mean and 90 per cent probability interval for the settlement curves (© with permission of GeoDelft).

ground risk evaluation, the resulting risk profile was considered as acceptable by the parties within the consortium. The last GeoQ step involved the mobilization of all relevant ground-related risk data towards the construction phase.

In line with the GeoQ process, the derived method of monitoring and back-calculation has been applied continuously during construction, which refined the ground characterization to an unusually high level during construction of the railway project. At the moment of completion of the construction, 3.5 years after the start of the work, the post-construction settlements were predicted at 0.2 m. The probability that the post-construction settlement is larger than 0.3 m is only 0.21 per cent, which is considered as very acceptable by both, the client and the contractor's consortium (Molendijk et al., 2003).

This case demonstrates the added value of a risk-driven approach during the detailed design phase of a project. In fact, the contractor's consortium took a substantial risk by bidding with a conventional and cheap settlement reduction method. They succeeded in reducing the major post-construction settlement risk by a combination of innovative calculation methods, the application of the probabilistic Monte Carlo analysis and a rigorous application of the observational method. It is in my perception an excellent example of how to do more with our ground data. In this case, not one additional detailed and advanced ground *investigation* was chosen as ground risk remediation measure, but additional detailed and advanced ground *interpretation* and *correlation* by using abundant monitoring data. The statement of no risk – no glory is applicable to ground risk management, as the dedicated geotechnical team of the contractor's consortium proved. This particular section was the only part of the entire Betuweroute railway project that was constructed well within budget and planning, in spite of the worst ground conditions of the entire project.

Summary

This chapter demonstrated the application of the GeoQ process in the design phase of construction projects, while focusing on the GeoQ step of *risk remediation*. Two approaches to support ground risk management during design were introduced and discussed. The first identifies and selects the most appropriate way of remediating foreseeable ground-related risk. General options are the reduction of the risk *causes*, the reduction of the risk *effects*, or a combination of both. Probabilistic approaches appear to be useful to obtain insight into risk cause reduction possibilities. However, the available techniques are not yet widely adopted, partly because of their highly specialist character. The combination of the observational method with fall-back scenarios proved to be useful for risk effect remediation. While the observational method largely depends on monitoring and back-analysis during construction, it should already be a due part of a flexible design, in order to be effective.

The risk-driven definition and application of ground investigations is the second approach to ground risk remediation during the design phase. Detailed and advanced ground data may be highly desired, in order to decide upon the most suitable risk remediation measures, while this information is obviously also required for conventional design activities, according to design standards and specifications. A six step model translates the causes of object-related risks to ground parameter level and serves as a basis for a truly risk-driven ground investigation. This model largely determines the type, quantity, and quality of the ground investigation. The model can be supported by probabilistic methods that balance the costs of ground investigations with the expected probability of reaching its objectives.

A few concise cases from different parts of the world illustrated the possible benefits of detailed and advanced ground investigations for risk remediation during design. Two detailed cases demonstrated the application of the described ground-risk remediation tools in practice. The first case has been demonstrating how the GeoQ process facilitates the design of a bored tunnel, below an existing railway founded on sand with a serious liquefaction risk. A detailed ground investigation with advanced techniques provided a sound basis for cost-effective risk remediation. The second case has been demonstrating how GeoQ contributes to reduce serious residual settlement risks during the final design of a railway project, with very unfavourable and soft ground conditions. A contractor's consortium won this design and construct project with a conventional and cheap method for settlement control and they succeeded in realizing the project within the challenging settlement margins by combining innovative approaches with proven techniques.

11 *GeoQ in the contracting phase*

Introduction

There is no construction without a contract. The drawing up of effective construction contracts has been developed towards a significant discipline within the construction industry. As construction projects become ever more complex, their contracts need more attention, which results in a steady increase in the number of pages over the years. The days with a contract of just several pages with a few attachments are gone, probably forever. Particularly if dealing with ground, transparent contracts with clear responsibilities for the parties involved are usually highly valued. Because of the inherent ground uncertainty with fuzzy, random and incomplete information, transparent ground-related contracting faces major challenges.

During the contracting phase, the client or owner selects a suitable contractor to realize the project. Contracting can take place in several phases of the project, for instance after the feasibility phase, the pre-design phase or the design phase. In this book the latter situation has been worked out, because this *Design-Bid-Build* (DBB) type of contract still represents the actual situation in many construction projects worldwide. For instance, the North American tunnelling industry applies DBB contracts for the majority of projects (Essex, 2003). In this book the DBB contracting approach is mentioned as the *conventional* way of contracting. There is an increasing trend, however, to involve the contractor earlier in the process of the construction project, after completion of the pre-design phase or even just after the feasibility phase. This type of *Design & Construct* (D&C) or *Design-Build* (DB) is considered a *modern contract* in this book. If maintenance also has been included, for instance during a 10-year or 20-year period after completion of construction, we reach the so-called *Design-Build-Maintenance* (DBM) sort of contract. Outsourcing the financing of the project to the contractor as well gives a *Design-Build-Finance-Maintenance* (DBFM) contract. Figure 11.1 shows the GeoQ ground risk management process to be well applicable to all of these contract types.

This chapter presents GeoQ as a facilitator of ground-related risk management during the contracting phase. The main GeoQ purpose is reaching the best contract for the construction project, seen from a ground risk management perspective and within the risk tolerance of the parties involved. This chapter builds forward on the answer to the last question in the introduction of Chapter 7: are we, our team members, our client and our other stakeholders willing to allocate each identified and classified ground-related risk to one

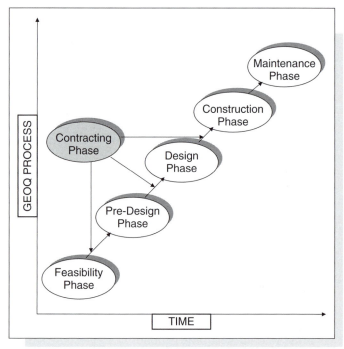

Figure 11.1: The contracting phase within the six GeoQ phases.

of the parties involved in the project? While assuming a clear and dedicated yes on this question, we explore in this chapter *how* to allocate ground-related risk in a transparent way, in spite of the inherent uncertainty of the ground. A paradox appears to arise here, but there is a way out.

Contractual risk allocation is a foundation for the last three GeoQ steps of *risk remediation*, *risk evaluation* and *risk mobilization*. In fact, effective risk remediation can only start after a clear risk allocation, because only the party explicitly responsible for a risk will be dedicated to control that risk. The three preceding GeoQ steps of *information gathering*, *risk identification* and *risk classification* are obviously indispensable to start any contractual (ground) risk allocation. We have to know our risks, including their seriousness, to be able to allocate them in some way or another.

This chapter starts with the introduction of the concept of *risk allocation* and describes its purpose and demonstrated added value. Next, the concept of *differing site conditions* is introduced and compared with the so-called *common law* approach. We will disclose the concept of differing site conditions as being fundamental to the allocation of ground-related risk. This reveals the need for a method to implement ground risk allocation in the construction practice by the concept of the *Geotechnical Baseline Report* (GBR), a promising method for ground risk allocation which has been successfully applied, since the 1990s, in several parts of the world. In the mean time, it is still subject to considerable debate. The GBR may develop to become one of the main tools within the GeoQ process, because it translates implicit ground uncertainty to explicit and controllable ground-related risk. The GBR's baseline concept, as well as the way to prepare a GBR is covered in relation to conventional and modern construction contracts. Three additional aspects, the *Dispute Review Board* (DRB), construction project *insurance* and the concept of *partnering* may further support ground-related risk management during contracting and the following project stages.

Finally, experiences with the application of ground risk allocation by the GBR are shared by a number of cases from the USA, the UK and The Netherlands. The application of the GBR during the resolution of disputes with differing ground conditions is worked out in more detail in Chapter 12. As usual, also this chapter ends with a summary.

Risk allocation and differing site conditions

The concept of risk allocation

A key success factor for effective risk management is the contractual *allocation* of risk arising from differing ground conditions. I cite David Hatem (1998) accordingly:

> Whatever the precise circumstances, the allocation and assumption (conscious or not) of heightened or intolerable risk for either the owner or the contractor on a subsurface project bears a direct relationship to the increased probabilities of disputes, claims, 'pathological' project relationships, and eventual litigation or arbitration.

The term *risk allocation* has been used a few times without presenting a definition of it. What exactly does it mean? Risk allocation implies an identified and classified risk becoming the explicit responsibility of one or more involved parties. For instance, the risk of unacceptable differential settlement becomes the responsibility of the contractor, as a result of allocating that risk to the contractor. If the risk occurs, the contractor bears the financial and other consequences. It is therefore in the contractor's interest to appraise, as soon as possible, the seriousness of the particular risk to be able to take timely appropriate risk remediating measures.

Why should we allocate risk? The answer is simple: risk will be managed and controlled when the risk has been explicitly allocated to one or more parties. The relevance of risk allocation seems to be time-independent. In the 1970s, Walter S. Douglas (1974) stated:

> What threatens the stability and financial security of the construction industry is not design, but the problems of distributing the risks inherent in the construction process among the owner, the construction contractor, and the architect and engineer...The industry cannot be healthy unless the risks are forthrightly recognised and acknowledged, and the various contracting parties assume under contract, without ambiguity, their respective parts of the risk.

This refers to the *manageability* of risk (Smith, 1996), each identified and classified risk needs to be assigned, consistent with widely accepted principles of risk allocation. According to the American Society of Civil Engineers (1980), these principles are:

1   Every identified risk has an associated and unavoidable cost that must be considered somewhere in the construction process

2   Risks should be the responsibility of those parties who are best able to control the risk, including bearing the costs and potential benefit

3   Many risks are best shared, with respect to their most cost-effective control.

The first principle is covered by the application of the GeoQ process, at least with respect to ground-related risk. The second principle explicitly advocates risk allocation, while the third principle assures the power of a mutual interest in risk management. These principles support the application of fair risk allocation that is likely to minimize the risk of litigation (Wildman, 2004). In addition, a reasonable risk allocation will reduce the likelihood of professional liability exposure for the involved design and construction professionals (Hatem, 1998), which may avoid a lot of unwanted hassle and cost for these professionals. There appears sufficient reason to explore risk allocation in more detail. I distinguish four main types of risk allocation:

1   *Unshared risk* for the client: the risk is entirely allocated to the client, who has to deal with the (cost) consequences, if the risk occurs

2   *Unshared risk* for the contractor: the risk is entirely allocated to the contractor, who has to deal with the (cost) consequences, if the risk occurs

3   *Completely shared* risk by both the client and the contractor: the risk is entirely allocated to both parties and together they have to agree their joint risk remediation measures, as well as which party is going to pay what portion of the cost, if the risk occurs

4 *Unshared risk*, part of the risk is the client's responsibility and the remaining part is the contractor's responsibility: the risk is partly allocated to the client and partly allocated to the contractors, by mutually agreed baselines.

This latter approach will be worked out in detail in the Geotechnical Baseline Report (GBR), later in this section. Figure 11.2 puts these four risk allocation options in one diagram.

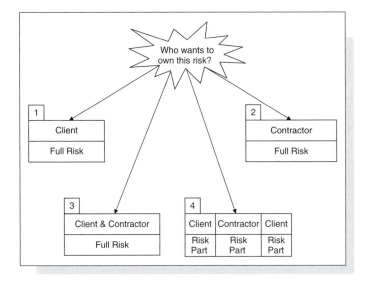

Figure 11.2: Risk allocation: four options.

Each identified and classified risk needs allocation to one of these four types of risk allocation to guarantee that each and every risk becomes explicitly owned by one or more parties, who will be more or less forced take responsibility for the risk. Risk allocation is not limited to ground-related risk. Many, if not all, remaining risk types can be explicitly allocated, such as performance-related risks and outside-influence sort of risk, including changes in governmental acts and adverse weather conditions (Smith, 1999). Anticipated advantages of explicit risk allocation for the client are:

- Lower bid prices, because contractors do not have to include contingencies in their bids for risk they cannot cost-effectively control; these risks are allocated to the client

- Lower total project costs, because of a reduction of disputes and claims, particularly concerning differing ground conditions

- Improved working relationships with the contractor, because it is clear which party is responsible for each identified risk.

Obviously, for the contractor clear benefits are also expected:

- Improved opportunities for competitive and innovative bids, within clear and pre-set risk responsibilities. The contractor is only responsible for risks that are manageable in some way. Innovative contractors may propose their clients allocate even more risks to them, because they have distinguishing solutions to control these risks. Therefore, contractors will be better able to bring their specific and unique competitive advantages to the market

- Higher profits, because the contractor bears only responsibility for a controllable set of risk. Contractors are be compensated for effectuated risk beyond their responsibility

- Improved working relationships with the client, because of the transparency of risk responsibility.

These advantages have been experienced in a number of projects, worldwide, and seem to pay off. A thorough study of two large transportation tunnel projects that have applied risk management by risk allocation revealed estimated cost savings between 4 and 22 per cent (Sperry, 1981). The Construction Industry Institute (CII) is a consortium of clients, contractors, engineers and universities, with its administrative basis at the University of Texas in Austin (Smith, 1996). According to CII research, well-performed risk management may reduce total project design and construction costs by 20 per cent, while the time required for design and construction can be reduced by 39 per cent (Construction Industry Institute, 1994). In addition, CII studies on contracting practices point to a 10:1 cost-benefit ratio of risk management in projects where contracting is accompanied by improved risk allocation (Smith, 1996).

In the following section we return to ground-related risk allocation. According to Smith (1996), the best controllable risk allocation is that of *contractual* risk allocation. How can we contractually allocate ground-related risk? We have to explore the concept of *differing site conditions* and its counter-approach of the so-called *common law*.

The concept of differing site conditions

Earlier we explored the relationship between ground conditions and failure cost in the construction industry. Unforeseen ground behaviour, particularly, is a main source of construction risk and associated problems. In the words of Heinz Brandl (2004): 'Experience tells us that the largest group of claims and disputes in the civil engineering field is in the ground'. In other words, *differing site conditions* of geological or man-made origin cause numerous disputes, claims, cost overruns, and delays in many projects.

However, despite our dedicated attempts to manage ground-related risk, actual site conditions will continue to differ from our expectations from time to time. Therefore, particularly from a contractual point of view, we need to be able to act effectively when we encounter these differing site conditions. Contractual allocation of ground risk appears to be an interesting option for reducing the adverse effects of differing site conditions. By translating ground information into contractual statements, the unfavourable effects of differing site conditions may dramatically reduce. However, first we have to acknowledge two entirely different schools of thinking about which party should bear the consequences of differing site conditions. Two common approaches, encountered in many parts of the world, are:

1   The application of the common law rule

2   The application of the Differing Site Conditions (DSC) clause.

The application of the common law rule implies that the contractor bears all risk of differing site conditions. In the USA, the common law rule applies in absence of a specific contractual statement. In Japan, however, the common law rule is non-existent. According to Tani (2001), it is common Japanese practice that clients issue additional work orders to overcome differing site condition problems of contractors. I am not sure whether this approach avoids any differing site condition problems in Japan.

A generally accepted escape route from the common law rule is the inclusion of a differing site conditions clause in the contract, which states that the contractor will be compensated for different site conditions, often under conditions (Abbott, 1998). The rationale behind the DSC clause is that ownership of the site implies ownership of its differing conditions as well (Hatem, 1998). The common law rule entirely neglects this automatism, by typically allocating all site condition risk to the contractor, as being the party to construct.

According to Gould (1995), the purpose of the DSC clause is simply to decrease the contingency that a contractor should include in the bid to anticipate differing ground conditions. Wildman (2004) motivates the need to escape out of the common law rule by avoiding adverse project situations, such as endless litigation over claims of unforeseen site conditions and contractors who went bankrupt. The latter were not able to complete the work, which is obviously a problem for the client as well. These experiences resulted in an increase in the application of the DSC clause, which is now almost standard in fixed fee construction contracts. At present, many standardized contract models, such as the international FIDIC and also the Dutch UAV-gc2005, include the DSC clause. In addition, agreements of the American Institute of Architects (AIA), the Federal Acquisition Regulations (FAR) and the Engineers Joint Contract Committee (EJCC) operate DSC clauses (Wildman, 2004). The concept is not new, as the first DSC clause was defined and approved in 1926, by the President of the USA himself. While acknowledging the concept of DSC being existent for many years, widely applied and being based on common sense, what can we exactly consider as differing site conditions? The

United States Federal Differing Site Conditions Clause of 1984 defines two general types of differing site conditions (Essex, 1997):

- Type I: subsurface or latent physical conditions, which differ materially from those indicated in the contract

- Type II: unknown physical conditions at the site, of an unusual nature, which differ materially from those encountered and generally recognized as inherent in the work of the character provided in the contract.

Type I differing site conditions deal in fact with *foreseen* risk, as a result of the inherent incomplete, random and fuzzy ground character. Both *information* risk and *interpretation* may result in Type I differing site conditions. Type II differing site conditions deal with *unforeseen risk*, which effectuates because the unknown cannot be known.

Why is it that, in spite of the DSC clause and its updates over time, disputes about differing site conditions have increased to a widely acknowledged almost unacceptably high level? For instance, the US National Committee on Tunnelling Technology reported low level subsurface site investigations leading to differing site conditions claims averaging 28 per cent of the contract price (Smith, 1996). Perhaps this situation occurs because Type I and II differing site conditions definitions only provide an *apparent* transparency about the owner's responsibility for unforeseen conditions. Guidelines on how to decide *what* ground conditions are reasonably 'materially different' and of an 'unusual nature' are still vague, if not absent. Obviously, we need a method to develop further the concept of different site conditions to a measurable and contractual statement. The baseline concept of the Geotechnical Baseline Report appears to be promising.

The Geotechnical Baseline Report

The baseline concept

By building forward on the concepts of risk allocation and differing site conditions, we need a practical method that allows us to allocate ground-related risk. We require an as objective as possible answer on the following question: *what* ground conditions are materially different? The baseline concept gives the answer to this question. The Technical Committee on Geotechnical Reports of the Underground Technology Research Council of the USA developed the concept of the *Geotechnical Baseline Report* (GBR) in the 1990s. It describes basically what type of site conditions are materially different by using so-called *baselines*. In the UK, these baselines are occasionally mentioned as *ground reference conditions* (British Tunnelling Society

2003; Construction Industry Research and Information Association, 1978), or *agreed model ground conditions* (Clayton, 2001). Also the term *benchmark* parameters can be encountered. In these cases the GBR is referred to as *Ground Reference Report, Agreed Model Conditions Report* or *Geotechnical Benchmark Report*. For reasons of consistency, I continue to use the terms baseline and GBR in this book.

The objective of the GBR is to provide *contractual definitions* of ground conditions. The GBR attempts to make explicit whether ground conditions are materially different and the differing site conditions clause is applicable, or not. So-called *key risk drivers* connect the baseline concept to the practice of ground risk management. In most cases, ground-related risk is caused, or at least highly accelerated, by one or a few specific ground parameters. A key risk driver is defined as a measurable parameter which largely determines or 'drives' a particular ground-related project risk. Ground-related parameters, such as the thickness and strength properties of ground layers, but also groundwater levels, concentrations of pollution and the presence of old foundation piles may all act as key risk drivers. *Baselines* are numerical threshold values of these key risk drivers and allocate risk.

How does the baseline concept work? Risks that relate to ground conditions, consistent with or less adverse than the agreed baseline value, are allocated to the contractor. Risks that result from more adverse values than defined by the baselines are allocated to the client. Risks are no free lunch and allocating (more) risk to the contractor is likely to increase the bid price. In this situation, the ground risk premium is paid in advance by the client to the contractor. On the other hand, the allocation of (more) risk to the client will reduce the initial bid price of the contractor, assuming competitive markets. By setting different baselines it will be possible to realize different risk profiles for the client and the contractor. Figure 11.3 presents the baseline mechanism of the risk of boulders that adversely affect construction activities, such as tunnel boring or driving of sheet-piled walls.

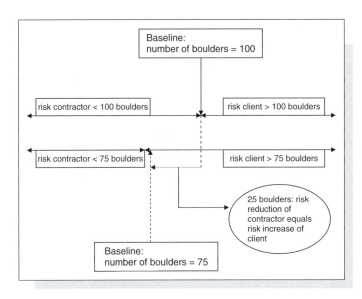

Figure 11.3: The baseline mechanism: allocating the boulder risk.

As Figure 11.3 demonstrates, allocating more boulder risk to the client equals less boulder risk for the contractor. In an open and competitive market, the contractor will have to reduce his bid price, as a result of his lower risk profile, because competitors are likely to do the same. With some creativity and practice, baselines can be defined for most of the geotechnical, geohydrological, geoenvironmental and man-made obstruction risks. Additionally, non-ground-related risks can be allocated by baselines as well, such as the traffic intensity from the number of passing cars per day, in the case of a DBM contract. Table 11.1 presents a number of examples of risk allocation by the baseline concept.

Ground-related risks, their related key risk drivers, as well as their baseline values are obviously project-specific. All of them depend largely upon the risk tolerance of the client and contractor, the type of project, and the anticipated ground conditions. I recognize a wide agreement on restricting baselines to physical parameters. So-called behavioural baselines depend, for instance, on the construction method applied and are therefore very difficult to anticipate, because different contractors usually apply different construction methods. Except for the risk of archaeological remains, all baselines, as presented in Figure 11.4, are of the physical type. The baselines for the archaeological remnants are not directly related to natural ground conditions as such, but to the presence of man-made obstructions. This risk depends largely on how effectively the archaeologist will do his job and is therefore considered beyond the responsibility of the contractor, when the delay and additional costs rise beyond the baselines.

A common pitfall of risk management in general and setting baselines in particular is growing out of control, as up to hundreds of identified risks apparently need to be allocated. In the process of risk allocation, we should realize that a large number of risks are already implicitly allocated to one of the parties within the (standard) type of contract used. This may also apply to a number of ground-related risks, while staying conscious of the inherent fuzziness of the different site conditions clause. Figure 11.4 presents the concept of the risk filter, which can be applied to keep the number of risks to be allocated as limited as allowable.

Table 11.1: The baseline concept: risk allocation by key risk drivers and baselines

Ground-related risk	Key risk driver	Baseline	Risk allocation	
			Contractor	Client
Unacceptable settlements	Thickness of clay layer	5 m	< 5 m	> 5 m
	Compressibility of clay layer	6 m²/kN	> 6 m²/kN	< 6 m²/kN
Groundwater inflow	Rock fracture permeability	10 cm/s	< 10 cm/s	> 10 cm/s
Polluted groundwater	Presence of hydrocarbons	200 ppm	< 200 ppm	> 200 ppm
Archaeological remains	Delay by excavations	2 months	< 2 months	> 2 months
	Additional cost	50 000 euros	< 50 000 euros	> 50 000 euros
Adverse weather	Temperature below −10°C	10 days	< 10 days	> 10 days

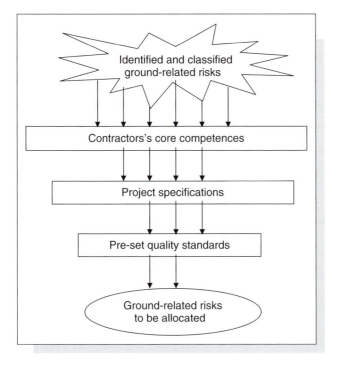

Figure 11.4: The concept of the risk filter for ground risk allocation.

According to Figure 11.4, risks that are clearly related to the core competencies of a contractor, the project specifications and the pre-set quality standards can be filtered out of the ground risk allocation process. Two of the original six risks in Figure 11.4 need finally to be allocated. The type and size of mesh of the risk filter depends largely on the characteristics of the project and the parties involved.

A GBR is meant to have a true contractual status, which is rather uncommon for ground-related information. Who of the readers is not familiar with the stamp 'for information only' on a factual, an interpretative or a geotechnical advice report?

However, while experts such as Morgenstern (2000a), Clayton (2001), Knill (2003), and Altabba, Einstein and Caspe (2004) advocate the importance of geotechnical risk management and the baseline approach, the concept is still surrounded with certain reservation by others. For instance, Brierly (1998) mentions the effects on the liability for professionals writing a GBR are as yet unknown. There is also no abundant case law associated with the use of baseline statements as part of the construction contract. In addition, Brierly (1998) mentions the debate on whether baselines should only reflect the most likely ground conditions to be expected, or should be used to allocate more or less of ground-related risk to either the client or the contractor, as I have demonstrated by the previous boulder example. Hatem (1996) questions the degree of detail of the baseline statement. He brings up an example of a baseline statement, which also concerns the number of boulders to anticipate. Let us now assume that the contractual baseline is set at 300 boulders, resulting from the client's low risk tolerance. Based on the available ground data, a best-guess assessment of 100 boulders would be reasonable in the particular glacial till deposit. Can the contractor claim any production-loss, after encountering only 50 boulders during the excavation activities, while he mobilized much slower excavation equipment to handle effectively 300 boulders? Figure 11.5 illustrates this baseline issue for the boulder risk during excavation.

This example illustrates the way towards clear baselines being still unpaved. In The Netherlands, where I joined a national committee to provide a Dutch GBR guideline, similar issues are subject to debate by both ground and contracts professionals.

On the other hand, if the GBR concept becomes embedded within an overall risk management framework, I foresee these issues will be solved in the near future. Nevertheless, we still face the

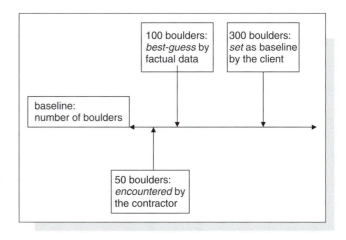

Figure 11.5: A baseline dilemma for boulder risk during excavation.

fuzziness and importance of the differing site conditions clause. For this reason I embrace the concept of the GBR, in spite of its actual and practical complications. The next section presents some guidelines on how to arrive at a GBR within the GeoQ process.

Preparing a Geotechnical Baseline Report

Straightforward guidelines for preparing a GBR are presented by the American Society of Civil Engineers (Essex, 1997). An update of these guidelines is scheduled to become available in 2006–2007. In recent years, I have encountered an increasing demand for additional 'know how' about the Geotechnical Baseline Report, in particular concerning its application within the GeoQ process.

The first three steps of the GeoQ process, *gathering of information*, *risk identification* and *risk classification* serve as the basis for the allocation of ground-related risk in a GBR. These GeoQ steps have to be done before entering the contractual phase of the construction project. The differences between the GBR, of which the primary purpose is the contractual allocation of ground-related risk, and other common ground-related reports need to be acknowledged. Typically related but different reports are the Geotechnical Data Report (GDR), the Geotechnical Interpretative Report (GIR) and the Geotechnical Design Summary Report (GDSR), as presented for instance by Brierly (1998) and Essex (1997). While of utmost relevance for applying the GeoQ steps of risk identification and risk classification, these reports should and can by no means replace these steps as well as a GBR.

In the contractual phase, any ground-related risk identified and classified needs to be (re)considered from a *contractual* rather than *technical* view. The contractual consequences of each of the significant ground-related risks need careful appraisal. The major question in this phase is who should own the ground-related risk and *why*. Obviously, the risk tolerance of the client plays a dominant role in answering this question, as well as other considerations, such as the possibility of reserving contingency budgets and the (local) political impact of differing ground conditions with severe effects on the community.

The degree of detail of the risks to be allocated depends largely on the type of contract. In a conventional contract, with a ready available project design, ground-related risks to be allocated require probably more detailed baselines than in a design and build type of contract, for which a set of functional specifications serves as the sole basis for any design yet to be made. In the latter situation, where totally different engineering solutions, varying from a bridge to a tunnel, have to be worked out, baselines may not need to be defined to the detailed level of ground parameters. More likely, sets of baselines are ground profiles and models, possibly combined with a number a factual index-properties. Furthermore, the presence of ground and groundwater pollution, as well as archaeological remains, can serve as baselines. However, also for this type of issue, the optimum approach has not yet been fully agreed by the professionals in the industry.

It has been widely agreed that normally the client provides a GBR as part of the bidding documents and the future contract. Therefore, the client has primarily to decide upon two major strategic options for the project with regard to risk and rewards:

1   To allocate (part of) the risk to the contractor, at an initial cost (the risk premium for the contractor) that is paid, independent whether the risk occurs or not

2   To allocate (part of) the risk to him- or herself and to provide a contingency budget, to pay only when the risk occurs.

In the latter option, the risk premium is, in fact, stored in a risk contingency budget, while recognizing that the risk premium and the risk contingency budget may be of different size. When the procurement procedure allows this flexibility, the contractor may provide alternatives for baselines. Therefore, an innovative contractor may be willing to accept more of a risk at the same price level than the competitors, because of better control of the risk of concern. In my opinion, ultimately, the innovative capabilities of the entire industry will be stimulated by this kind of risk allocation mechanism.

Another debate, already ongoing for decades, concerns the *status* and *degree of disclosure* of the available ground data to the bidding parties. In other words, what type, quantity and quality of information should the client supply in the procurement process and what is the legal status of that information? Occasionally, clients choose the option 'for information only', attempting to avoid any responsibility for the effects of possible misunderstanding or even omissions in their own data. In view of a transparent and fair bidding process, supported by well-understood risk management principles by both the client and the bidding contractors, I advocate *full disclosure* of all ground-related data. Experience shows such full disclosure contributing to risk reduction for both the client and the contractor, as well as to lower project costs (Munfah et al., 2004; Smith, 1996). In addition, courts state over and over again that the client has an *obligation* to make the contractor at least aware of all information that may affect the contractors' activities in terms of costs and time (Brierly, 1998).

The way to set baselines has been described earlier. Many ground-related baselines can be defined using existing project information data, as provided by the previous project phases but, in the case of very critical risks and related baselines, additional, detailed and advanced ground investigation data may be required. Similar to the concept of partial safety factors in geotechnical engineering, a surplus or reduction can be applied to geotechnical baseline parameters in order to allocate more or less contractual risk to the contractor. However, we should remember the discussion in the previous section concerning the baseline example of the boulder risk. Setting baselines beyond the most likely ground conditions is not without debate in many geotechnical and contractual communities. Nevertheless, the following criteria can be considered during the process of setting the probably most appropriate baseline parameters:

- The anticipated geological variation or heterogeneity on the site

- The degree of impact of the proposed key risk driver on its related risk

- The severity of the risk, expressed by its likelihood of occurrence, multiplied by the anticipated impact of the risk

- The preferred risk tolerance of the client, contractors and any third parties involved, as well as the associated cost of their (lack of) risk tolerance.

While appraising these criteria in view of the complexity of most construction projects, the reader may agree that selecting key risk drivers and their baselines is, like architecture and parts of engineering, more of an art than a science. However, state-of-the-art ground engineering is not yet adequately prepared to assess the *contractual* implications of baselines, in addition to their *technical* implications. This situation opens a window of opportunity for the ground-related disciplines to develop more advanced methods and software to facilitate the baseline selection process. Methodologies and means to assess easily the financial impact of different baseline proposals on the project results would make a major step forward.

Nevertheless, we should not feel discouraged by this lack of technology. For many years, modern software allows provision all kinds of ground-related sensitivity analyses, which can highly facilitate any decision-making process for the most suitable baselines.

In this context we meet again the decisive role of the people factor in the adoption of an innovative approach. With a high degree of individual and team willingness and true dedication to reduce the adverse effects of differing site conditions, the application of a GBR may provide major improvements in many projects. There is no chicken without an egg and we have no eggs without a chicken. The GBR will never become common practice without clear communication of its proven benefits. We will, however, not realize these benefits without increasing our GBR experience. The cost of a GBR should not obstruct its application compared with the costs of one single claim about differing site conditions. It is up to clients and their consultants to decide upon the degree of effort and associated costs to invest in a GBR. As I will demonstrate in the case section of this chapter, there are already favourable benefits of the GBR in several countries, which fade away its rather limited cost. Fortunately, I do not stand alone with these statements. The British Tunnelling Society (BTS) and the Association of British Insurers (ABI), which represent insurers and re-insurers on the London-based insurance market, issued their *The Joint Code of Practice for Risk Management* (Mining Communications, 2004). The GBR concept is entirely adopted in this code (British Tunnelling Society, 2003).

Finally, a key success factor of the application of GeoQ in the contracting phase of a construction project is completely to integrate it into the project's *procurement system*. Risk allocation needs to be embedded in procurement. This requires a more professional approach towards ground-related risk in the contracting phase, for instance by avoiding playing 'the ostrich game'. Contracting on lowest price only becomes simply impossible if different bidding parties have bids with explicitly different ground risk profiles. An evaluation of these different risk profiles by clear criteria is required, which is in fact the *risk evaluation* step of the GeoQ process during contracting. A further in-depth exploration of this procurement issue goes beyond the focus of this book. However, by now many readers may agree that the combination of procurement and risk allocation needs adequate attention in the years to come, in order to apply ground risk management effectively and efficiently in the contracting phase of our projects. In my opinion it is an indispensable element of the financial success that we chase by the application of the GeoQ ground risk management process.

Ground risk management during contracting

We may conclude that the GBR is a promising means to specify the inherently fuzzy differing site conditions clause. This aims to minimize the likelihood of money- and time-wasting disputes and litigation. This section introduces three additional risk remediation approaches that support ground risk management in the contracting phase of construction. The purpose of this section is to demonstrate how these measures may *contribute* to the application of a GBR within the GeoQ ground risk management process. It is certainly not the intention that these measures *replace* the GBR and the GeoQ process.

The Dispute Review Board

The *Dispute Review Board* (DRB) is a proven concept to avoid and resolve ground-related disputes and litigation in a cost-effective way. Research by Matyas et al. (1996) indicates a remarkably high resolution rate of almost 100 per cent. A DRB generally consists of three persons, one representing the client, one representing the contractor and a third who acts as chairman and will be selected by the other two members (Munfah et al., 2004). The members of the DRB inform themselves on a regular basis about the progress of the project in relation to the actual ground conditions encountered. Visiting the construction site at a regular time interval is common for a DRB. While the recommendations of DRBs are not legally binding, their findings are not likely to be reversed in litigation, due to usually thorough knowledge of both technical and contractual aspects of the DRB members (Caspe, 1998). According to Munfah et al. (2004), widely acknowledged benefits of the DRB are:

- Lower bid prices

- Improved communication and less acrimony on the project site

- More timely and cost-effective resolutions

- Fewer claims.

The DRB can, therefore, be considered as some sort of overall risk remediation measure for reaching the main project objectives. The DRB is preferably already established in the project's contracting phase, which distinguishes it from other so-called *Alternative Dispute Resolution* (ADR) methods. The latter aims to solve disputes when they are there already, while the DRB aims to avoid serious disputes. ADR methods concerning the construction phase are further described in Chapter 12.

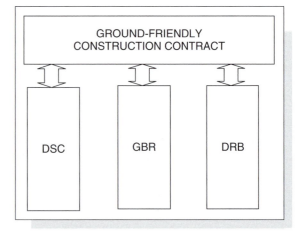

Figure 11.6: The three pillars of a ground-friendly construction contract.

The DRB can be considered as one of the three main pillars for a cost-effective construction contract, with adequate attention to ground-related risk, as illustrated in Figure 11.6. The Differing Site Conditions (DSC) clause and the Geotechnical Baseline Report (GBR) form the other pillars.

In the 1990s, the application of the DRB increased dramatically in the USA. Countries outside the USA where the DRB concept has been applied include China, India, Honduras, South Africa, Canada, The Netherlands, France and the UK. In the latter two countries, the Channel Tunnel project, connecting the UK and France, used a DRB. Several international organizations embrace the DRB concept as well. Matyas et al. (1996) mention for instance:

- The World Bank

- United Nations Commission on International Trade Law (UNICITRAL)

- Fédération Internationale Des Ingénieurs Conseils (FIDIC)

- International Chamber of Commerce (ICC)

- UK Institution of Civil Engineers (ICE)

- Engineering Advancement Association of Japan (ENAA).

It is beyond the scope of this book to explore the DRB concept in more detail. For instance, the *Construction Dispute Review Board Manual* by Matyas et al. (1996) provides a wealth of practical recommendations and experiences. Because of its inherent transparency, applying the GeoQ ground risk management process may further increase the effectiveness of a DRB. The combination of the DRB, DSC and GBR forms the core of the GeoQ process in the contracting phase and I foresee a bright future for the joint application of these concepts, with mutual benefits for all parties involved.

Insurance and ground risk management

The well-known risks remediation measure of *insurance* is usually applied for risk with a low likelihood of occurrence and (very) high effects. We insure ourselves against car accidents and our houses against fire. The insurance of construction projects, in particular those such as tunnels with a major ground risk exposure, is a specialized discipline. For reasons of completeness I want to address the insurance issue, however, in a concise scope.

Ground risk is often in some way implicitly part of project insurance. Typically, clients require contractors to carry a comprehensive general liability insurance that should protect the client's, the engineer's and any third parties' interests. For instance, a third party needs to be financially compensated in the case of property damage due to unexpected ground deformations during excavation. Wildman (2004) advises the client to purchase a client controlled insurance policy, which includes all parties involved in the construction project, in order to minimize the dispute and litigation potential. Gilmartin (1998) presents the following risk exposures that need to be covered by insurance in construction projects:

- Professional liability

- Commercial general liability

- Workers' compensation and employers' liability

- Builders' risk

- Construction equipment

- Environmental issues and the contractor's pollution liability.

It is considered beyond the scope of this book to explore all these insurance types in more detail. For instance, Gilmartin (1998) provides a clear introduction to the presented types of insurance.

Due to the increasing complexity of construction projects, insurance premiums have increased. In The Netherlands, it has become almost impossible to obtain an insurance, at a reasonable premium, for the rather high risk construction projects such as horizontal directional borings. This situation is an additional driver for structured ground risk management, either for convincing insurance companies about reasonable ground risk control, or for realizing the project without insurance, but within an acceptable overall risk profile. In recall in this respect the risk-driven *Joint Code of Practice for Risk Management of Tunnel Works in the UK* has been *initiated* by the Association of British Insurers (ABI) (Mining Communications, 2004). Box 11.1 explains why.

The option to insure ground-related risk should be seriously considered as part of the risk allocation process in the contracting phase of any project, while we need to acknowledge that *insuring* a risk is not the same as *eliminating* a risk. It is typically a *risk effect* remediation measure, in which the effects are financially compensated after they occurred. Therefore, we deliberately need to consider with our project team and client whether we can allow a risk to occur and be financially compensated afterwards, or that it is more favourable to implement *risk cause* remediation measures, in order to reduce the risk probability of occurrence. Again, illustrated by Box 11.2, clear and uniform answers about whether and which ground-related risks are better to be insured are non-existent.

Box 11.1   Why British insurers advocate risk management

Based on data from insurer Munich Re, Knights (2005) presented 15 tunnel projects over the period 1994–2004, which all faced major ground-related problems with financial losses, in total, of more than 500 million US dollars. These tunnel problems occurred in Europe, North America and the Far East. One striking example is a 10 km and 60 million pound tunnel project in Hull, the UK, which collapsed over 150 m. The insurance claim for reinstatement was some 42 million pounds, which implies more than 40 times the original scheduled costs for that 150 m tunnel length. These figures explain why the British insurers initiated the mentioned risk-driven code of practice.

Like the example of Box 11.2, the GeoQ process steps during the contracting phase may challenge the conventional ways of project insurance, to arrive at more cost-effective risk remediation measures within an acceptable risk profile of the project. Innovative partnering agreements may help in this respect.

Box 11.2   Risk management replaces insurance

The North-South Metroline is a major underground project which intersects the historical centre of Amsterdam by a bored tunnel with a number of deep stations. From the earliest beginning, risk management has been a major issue in this project. Very comprehensive risk remediation by the observational method and detailed monitoring has been applied. The public client's project team considered therefore it permissable to omit insurance against third party liability. Instead, the reserved insurance premium served as a contingency budget to allow payment to third parties in case of any damage because of the construction activities.

From conventional contracts to partnering contracts?

Many construction project risks, including the well-known ground-related ones, are principally independent of the type of contract. However, the *allocation* and *responsibility* of these risks are highly dependent on the type of contract. For instance, some standard types of contract in the UK, like the Joint Contracts Tribunal (JCT) and the New Engineering Contract (NEC) allow several contractual purposes with different risk allocations (Edwards, 1995). The most suitable type of contract for a particular construction project depends on many factors, including the risk attitude of the client and the type and complexity of the project, like a surface or subsurface project (Brierly and Hatem, 2002). Even contract specialists occasionally do not agree about the most suitable contract for a project. Their contract preference can be rather subjective and appears to be a matter of (risk) taste. We may recall here the influence of the people factor and the inherently different perceptions of the project stakeholders.

As for instance indicated by Essex (2003) and Wildman (2004), most tunnel projects follow still the conventional type of DBB contracts, which separates design from construction. If a problem occurs during construction, such as unexpected groundwater inflow in a tunnel or building pit, normally the following question arises: 'Is it a design problem or is it a construction problem?' In the first situation, in the case of a design problem, the client or engineer is responsible, while the contractor bears responsibility in the second situation. Ground conditions, by differing site conditions, often play a dominant role in differentiating a design problem from a construction problem. If the parties involved can quickly agree on whether the actual site conditions are differing or not, the responsibility question is much easier to answer. As a matter of fact, also in conventional contracts, the GBR and DRB can save a lot of money and time by classifying ground conditions as differing or not.

Wildman (2004) relates a rather risk adverse attitude of today's design community to the inherent uncertainty of subsurface construction. Clients may have to defend their engineers against claims of contractors, for instance in cases of differing site conditions. From a client's perspective, this unfavourable perspective can be a main driver to advocate the modern D&C or DB type of contracts, in which there is one single responsibility for engineering and construction. Other benefits are an increased opportunity for *innovation*, because design and construction professionals work together in one team. In case of shared performance goals, these innovations are expected to result in savings in cost and time (Essex, 2003). The number of innovative contracts in construction projects tends to increase, which appears to be a global trend. In fact, James P. Gould highlighted this trend during The Twenty-Sixth Karl Terzaghi Lecture in 1995: 'The overriding purpose of improved contracting practices is to provide a constructed product at least cost for the owner, with an appropriate profit to both the sponsor and contractor'. He added to this: 'Success of this process depends largely on the performance of the geotechnical engineer'.

In addition, the concept of the *target price contract* is evolving, a special type of D&C contract, in which a pre-set price for a fixed scope of work has been agreed. The contractor will gain an incentive fee that increases if the contractor operates well below the target price. If the costs of the contractor exceed the target price, then the fee is reduced. The British Airport Authority is including this type of price mechanism in its contracting approach (Finch and Patterson, 2003). The chasing

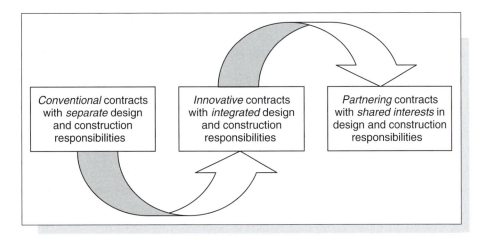

Figure 11.7: A ground-friendly contractual evolution?

of mutual benefits by this type of contracting strategy evolves into some sort of partnering. Do we have to move from conventional contracts, via innovative D&C types of contract towards partnering or alliance contracts, as the ultimate cost-effective and ground-friendly type of contract in construction? Figure 11.7 illustrates such an evolution.

This contractual evolution is interesting from a risk management viewpoint, because it aligns, to an increasing degree, the interests of the client and contractor in joint ground risk management. This may act as a driver for cost-effective reduction of the adverse effects of differing site conditions. Particularly in the offshore oil and gas industry, *relationship-based contracts*, like partnering agreements, proved to have substantial performance increase and cost improvement. Blockley and Godfrey (2002) describe partnering as a structured management approach to facilitate teamwork across contractual boundaries. The high degree of interdependence between various organizations in the supply chain of the offshore industry makes it especially suitable for partnering.

While the construction industry can be characterized by a similar high interdependency in the supply chain, for instance, by the many different suppliers and subcontractors in most of the projects, the concept of partnering is relatively new. In 1989, the US Army Corps of Engineers (COE) brought partnering to the industry (Neff, 1998), with the following definition:

> Partnering is the creation of an owner-contractor relationship, that promotes the achievement of mutually beneficial goals. It involves an agreement in principle to share risks involved in completing the project, and to establish and promote a nurturing partnership environment (US Army Corps Of Engineers, 1989).

This definition is more or less similar to the one in the UK, which adds *dispute resolution* explicitly in its partnering definition (Construction Task Force, 1998).

According to Neff (1998), partnering does *not* involve a contractual agreement, nor does it create any legally enforceable rights or duties.

The partnering process may start *after* contract award (Smith, 1996), which resulted in a success story in The Netherlands, as described by Box 11.3.

Box 11.3   Partnering raises 25 million euros

Around the year 2000, after negotiations of nearly one year, the client and contractor's consortium of a part of the Dutch Betuweroute railway project transformed their D&C contract towards a contractual partnering agreement, the first one in The Netherlands in such a type of construction project. Risk sharing rather than risk allocation to separate parties, including ground-related risks, was one of the pillars of this contract. The total costs savings in this project were some 25 million euros, which is about 10 per cent of the budget at contract award. Like risk, the client and the contractors also shared the benefits of these savings on a 50:50 basis.

Remarkably, there has been no legal escape built into the contract to get out of the partnering contract. The client and contractors were contracted to each other, which has been proven to work effectively. The people factor ruled and succeeded as well. As indicated by Wuite (2005), both representatives of the client and the contractor worked effectively and pleasantly together. They preferred their partnership above to conventional contractual fights.

It could be even more effective to establish the partnering arrangement *during* the contracting process itself in order to gain a (more) legal status, for instance, with regard to a shared and contractual allocation of ground-related risk. Munfah et al. (2004) describe a formal partnering provision that has been be incorporated into contracts by the New York Metropolitan Transport Authority for their 6300 million US dollar East Side Access underground project.

Blockley and Godfrey (2000) advise building relationships and sharing mutual values, needs and objectives even *before* contracts are signed. Partnering candidates may very well start with a team-based risk brainstorming session, as described earlier, in which issues of concern, like different risk perceptions, as well as potential challenges for value engineering are raised and addressed. The results of the GeoQ process, as performed in the project phases preceding contracting, may typically serve as a useful foundation for the partnering process.

Partnering can be considered as contributing to the people factor and the soft systems in construction projects. As described in Box 11.3, the effects of partnering depend largely on a positive attitude of the representatives of the parties involved.

Shared interests of clients and the contractors will prove to be the main drivers to work together. Therefore, any shared goals, objectives and risk demand crystal-clear communication while establishing any partnering relationship.

The concept of partnering rises preferably beyond project-level to industry-level. The need for it can be retrieved from a statement made by Abramson et al. (2002):

> The future of underground development in North America depends on our ability to accurately forecast underground tunnelling conditions, estimate costs, and complete tunnels within those estimates. The contractor's collecting and sharing these performance data with equipment suppliers and tunnel owners should assist in minimising misunderstandings relating to the development future of underground infrastructure development.

This seems to provide a clear view on the necessity to mobilize the entire supply chain within the industry for joint innovation, which supports further development of not only underground infrastructure but construction activities in general.

A thorough application of the GeoQ process may largely contribute the necessary trust building within any partnering process, because it demands full disclosure by risk registers and all other relevant project information. The National Association of Dutch Contractors supports a shared risk allocation of, among others, ground-related risk by partnership types of contracts (Koenen, 2004). Matyas et al. (1996) perceive the positive attitudes, as fostered by partnering, as fully compatible with the concept and intentions of the DRB. All of these developments tend to point in the direction of cost savings for clients and increased profitability for contractors. The following section presents a number of cases demonstrating how good intentions can be transformed to the pursued concrete and positive results.

Case studies

Despite its promising concept, since starting in the USA in the 1990s, contractual ground risk allocation by the GBR appears to remain in a rather early phase of development. Otherwise, I meet a lot of enthusiasm with construction professionals in a variety of countries, such as the USA, the UK, Norway, Germany, France, Greece, China, Japan and The Netherlands.

The number of GBR-related case studies describing the contracting phase within a risk-driven framework proves to be rather limited. For instance, the proceedings of the sixteenth International Conference on Soil Mechanics and Geotechnical Engineering (ICSMGE) held in Osaka in 2005, includes some 600 papers. However, less than 1 per cent of these papers explicitly concerns ground risk management and none of them mentions the GBR or a similar concept. In this respect, professor Frans Barends of the Delft University of Technology 'saved' the conference by his Terzaghi Oration 2005, by one of his conclusions:

> Transparency of the uncertainty in geotechnical works will facilitate proper risk assessment
> and risk sharing by use of appropriate forms of contracts, avoiding inflexibility, indecision
> and time overrun (Barends, 2005).

Because of this actual situation, I present in the following a number of concise case studies from the USA, the UK and The Netherlands, to provide an overview of today's worldwide GBR-related experience in the contracting phase of construction projects.

The USA and contractual ground risk allocation

Matyas et al. (1996) provide some general but promising GBR experience. For instance, since the Washington Metropolitan Area Transit Authority (WMATA) has introduced the GBR, the number of site condition claims and subsequent financial awards has decreased significantly. All tunnel designers engaged by the WMATA have since accepted the concept. Other clients, who incorporated the GBR concept in their projects are, and not limited to, the Alaska Power Authority, the cities of Los Angeles and Honolulu, the municipalities of Anchorage and Seattle and the Colorado Department of Highways. This list expresses a wide variety of GBR-applying clients from all over the USA. Reports from clients, engineers and contractors seem to judge the GBR concept as consistently favourable. Some credited the GBR for saving millions of dollars because of the avoidance of claims and litigation.

Wildman (2004) describes the application of the DSC clause for the Milwaukee Metropolitan Sewerage Tunnel Project. The contractor encountered materially different ground conditions of large water inflows and poor rock support during the project execution, compared with the ground conditions as represented in the contract documents. While the GBR is not explicitly mentioned in this paper, the engineer of the owner granted the differing site conditions status to the contractor's construction problems. The owner ended up paying 166 million US dollars to the contractor, almost four times the contractor's original contract price of 46 million US dollars. This demonstrates the impact that differing site conditions can have on a project. Wildman suggests considering a D&C contract for similar projects, to establish a single point responsibility to the designer-builder, which may protect the client against significant design-error claims by the contractor.

Munfah et al. (2004) presented a very large scale GBR-case. The earlier mentioned 6300 million US dollar East Side Access Underground Project is the largest transportation project ever undertaken in New York City. A continuous risk management process, including of risk identification, risk quantification, risk response development and risk response monitoring was started at the early stages of the project, which appears in line with the GeoQ process as described is this book. Due to the high risk exposure of this typical greyfield project, with numerous adjacent and overlying structures, the client judged the conventional design-bid-build process as most appropriate. The contract documents provided full-disclosure of ground data, by including a GDR, GIR and GBR. The latter includes quantitative values for selected ground conditions that are expected to have a great impact on construction. These values are established by *technical* interpretation, while considering the *commercial* implications for ground-related risk allocation and sharing between the client and the contractor. The client decided to take some additional portion of risk, in exchange for a lower initial bid. Here the baseline mechanism, as discussed with a balance between risk allocation and initial bid prices, appears to be applied in practice. Munfah et al. (2004) present the following anticipated advantages by the application of the GBR in this mega-project:

- Ease of administration of contractual clauses

- Unambiguous determination of entitlement

- Clear basis of contractor's bid

- Clear allocation of risk between client and contractor.

The award of the first tunnelling contract of this project was scheduled to be early 2004, while the entire project is planned for completion by 2012.

The UK and contractual ground risk allocation

The UK practice on ground reference conditions, the British equivalent of the baseline approach, was initiated in the 1970s by the Construction Industry Research and Information Association (CIRIA). However, it has not yet been widely adopted by the UK construction industry. In 2002, the late Sir John Knill (2003) stated in the First Hans Cloos Lecture : 'It is now regarded as good practice for as much information to be provided to a contractor as possible as part of the contract documentation, within the framework of carefully devised baseline conditions'. The British Tunnelling Society (BTS), together with the Association of British Insurers (ABI), integrated the statement of Knill in their The Joint Code of Practice for Risk Management in the UK, by the adoption of a truly risk-driven and staged project approach, which include the GBR-approach (British Tunnelling Society, 2003).

Rigby (1999) provides a remarkable case study, in which not only the GBR but even the entire GeoQ process is reflected. Contrary to the major projects of the previous cases, the John Pier Tunnel Project in the north-west of the UK is a much smaller activity. It demonstrates the suitability of GeoQ-related ground risk management processes, including the application of a GBR, for rather small projects as well.

The project described by Rigby (1999) includes a 2 km long, 2.9 m diameter, tunnel in challenging ground conditions. The site included old mine shafts, silts that are susceptible to liquefaction and glacial deposits with very strong boulders showing rock strengths up to 350 MPa. These ground conditions were retrieved from an initial desk study, which was followed by several phased site investigations. A longitudinal section along the selected tunnel route, showing the anticipated ground conditions, served as the baseline conditions. Ground-related risks were identified, classified, allocated and partly shared between the client and the contractor. A New Engineering Contract (NEC) type of contract has been selected for its flexibility in terms of the level of design responsibility and pricing methods. A partnering arrangement has been included, as well as incentives to align the objectives of the client and the contractor. During the tunnel construction process, the encountered ground conditions were compared with those as agreed by the baselines. All of the cost savings that resulted from risk mitigation were shared on an agreed basis between the two parties. This entirely risk-driven project proved to be very successful, with a 20 per cent faster completion than scheduled and 10 per cent cost savings. In my view, this case clearly demonstrates the benefits of a GeoQ-like approach, including its contractual consequences by the GBR application, for a project with a relatively modest size.

The Netherlands and contractual ground risk allocation

Since its introduction, the GBR has been applied in some ten projects in the period 2000–2005 in The Netherlands. Besides the cases presented in the Boxes 11.2 and 11.3, a bored tunnel with deep stations and a railway project, the Dutch experiences include a number of highway extension projects and some waterworks. These projects varied in size, from rather small to very large. In addition, Jansen (2001) reveals the implicit application of GBR-alike approaches in a few major D&C projects over the last years. All of the clients of these projects were publicly owned.

A few Dutch workshops in the period 2001–2003 revealed optimism about the opportunities and benefits of the GBR. A DSC is standard in most Dutch contracts, which allows application of the GBR concept for both conventional and modern types of contract. However, some hesitation for a real GBR breakthrough is still there and a number of barriers may be heard of:

- The GBR concept is still rather unknown in large parts of the Dutch construction industry

- The advantages of the GBR are not yet made explicit by the Dutch practice

- The GBR is not yet completely fit-for-purpose towards the Dutch design, contracting and construction practice.

To raise more attention, for ground risk management in general and the application of the GBR in the contracting phase in particular, a number of papers, courses and lectures have been provided (van Staveren, 2001a, 2004; van Staveren and Peters, 2004; van Staveren and Knoeff, 2004; Herbschleb et al., 2001; van Staveren and Litjens, 2001). A major step was the start of an industry-wide technical committee in 2004, chaired by the Dutch Ministry of Public Works. The ultimate task of the committee is to prepare a guideline for the application of the GBR in the Dutch contracting and construction practice. The awaited benefits of the GBR in the Dutch construction practice are mainly similar to those pursued in the USA and the UK:

- Reduction and more cost-effective resolution of differing site conditions claims

- Providing more objective and uniform insight into the project's ground conditions

- Improving communication between the project parties about ground-related risks

- Supporting innovative and cost-effective engineering and construction within a reasonable risk profile for the parties involved

- Supporting value engineering practices for delivering cost-optimizations

- Supporting the life cycle cost approach for reducing total cost of ownership.

By being thoroughly embedded in the GeoQ ground risk management process, the GBR contributes to the increasing risk-aware climate in the Dutch construction industry. The Dutch GBR guideline will be available in 2006. From that moment it is entirely up to the willingness and dedication of the change agents in the Dutch industry to apply the GeoQ-GBR combination in their practices, in order to materialize the anticipated and appreciated benefits.

Summary

Still, there is no construction without a contract. This chapter demonstrated the application of the GeoQ process in the contracting phase of construction. The willingness to *allocate* each identified and classified ground-related risk to one of the parties involved is a key factor during contracting. This contractual risk allocation serves the last three GeoQ steps of *risk remediation, risk evaluation* and *risk mobilization*. As a matter of fact, effective risk remediation can only start after a clear risk allocation, because only the party explicitly responsible for a risk will be dedicated to control that risk effectively.

With regard to ground-related risk allocation, the concept of *differing site conditions* (DSC) has been compared with its opposite, the so-called *common law* approach. The DSC appears to be the favourite with regard to the anticipated benefits, for both the client and the contractor, in either conventional or modern types of contracts. The DSC needs a contractual specification by the *Geotechnical Baseline Report* (GBR). By making an explicit relationship between geotechnical risk, its risk driver and the baseline of that risk driver, each identified and classified ground risk can be clearly allocated to a party, within well-defined boundaries. In other words, the GBR makes *explicit* whether ground conditions are materially different from the anticipated ground conditions, or not. The inclusion of the GBR in a structured risk management framework, such as the GeoQ process, appears to be rather new.

Three proven risk remediating approaches have been introduced: the *Dispute Review Board* (DRB), construction project *insurance* and the concept of *partnering*. These approaches both fit and strengthen the GeoQ ground risk management framework.

Experiences with the application of ground risk allocation by the GBR within GeoQ-like risk management processes have been presented by a number of cases from the USA, the UK and The Netherlands. These cases demonstrated promising results in terms of savings in cost and time.

12 GeoQ in the construction phase

Introduction

In construction, the proof of the pudding is in the eating. After occasionally several years of preparation, the core of the construction process will start: construction itself. During this phase the fit-for-purpose of design, with all risk remediating measures taken, will be tested in practice. In spite of all preparations that have been made, surprises during construction remain inevitable. We will encounter pleasant surprises, like ground pollution that demonstrates to be less severe during the clean-up process at a brownfield site, as well as unpleasant surprises, such as hitting a buried water pipeline at a greyfield site, which was clearly not indicated on the drawings. However rigorous our preparations have been, uncertainty will accompany us during the construction process.

The same applies for the GeoQ ground risk management process. While more and earlier attention to ground risk management importantly reduces the number and severity of foreseeable risk, a residual uncertainty remains. In my view, effective ground risk management is more about dealing with ground uncertainty, in full awareness, than totally eliminating ground uncertainty. The latter is an unrealistic dream scenario, in particular during construction. Figure 12.1 presents the construction phase in the entire GeoQ process.

In the case of a conventional type of construction project, construction follows after the contracting phase. Otherwise, construction succeeds the design phase. By building on the examples in the introduction of the previous chapters: now the bored tunnel actually has to be bored. GeoQ aims to provide structured ground risk management during the entire construction process, helping to realize the most favourable construction result, within the project specifications, the agreed risk profiles and the ground conditions to be encountered. Ideally, also in this phase all of the six GeoQ steps are performed.

The result of the construction phase is a well-completed construction project. Both the construction process and the construction result will have either a positive or a negative impact on the reputation of the contractor from the viewpoint of the many project's stakeholders. For instance, regarding a greyfield project in a city centre, the level to which the contractor is able to reduce the hindrance for the public can have a massive impact on the contractor's reputation with the public and the (government) client.

This chapter starts by

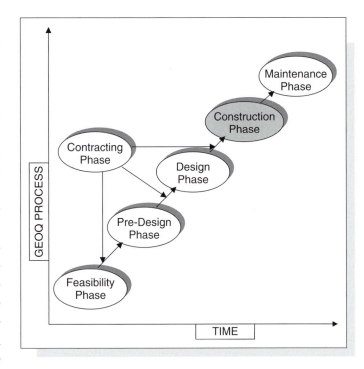

Figure 12.1: The construction phase within the six GeoQ phases.

introducing and discussing two methods to support ground risk management in this particular construction phase. These methods belong to the fourth and fifth GeoQ process steps of *risk remediation* and *risk evaluation*. These steps are vital during construction to arrive at an as smooth as possible construction process.

The first method revisits the observational method from a risk management perspective during construction. I predict a revival of this approach, because of today's wealth of possibilities for on-line monitoring and almost real-time back-analyses. This concept of back-analysis, the evaluation of design calculations by the application of monitoring data is considered an integral part of the observational method in this book.

The second method for ground risk remediation during construction covers the role of the GeoQ process, including the concept of the Geotechnical Baseline Report (GBR), when construction faces differing site conditions. A number of proven ground dispute resolution options are presented, because the GeoQ process may highly facilitate them. Finally, two case studies will be presented in which the GeoQ process played a dominant role. As usual, the chapter ends with a summary.

Ground risk remediation during construction

The observational method revisited

While Ralph Peck made the *observational method* widely known by his Ninth Rankine Lecture in 1969, the founding father of soil mechanics, Karl Terzaghi (Figure 12.2), introduced and used the method during the 1930s.

The following words of Peck (1969) concerning the observational method during the Rankine Lecture remain relevant:

> In spite of the limitations, the potential for savings of time and money without sacrifice of safety is so great that every engineer who deals with applied soil mechanics needs to be informed of its principal features.

Figure 12.2: Karl Terzaghi: (1883–1963) founding father of soil mechanics.

In addition, I expect a *revival* of the observational method in the years to come, due to the following three developments over the last few years:

1 The increasing tendency to integrate design and construction into one contract

2 The increasing availability of cost-effective monitoring techniques, in combination with on-line information and communications technology (ICT)

3 The increasing demand for effective ground risk management.

These trends are favourable to a bright future for the concept of the observational method as one of the main risk remediation measures during construction. Basically, the observational method monitors ground behaviour during construction. Appropriate fall-back scenarios can challenge the apparent limits of ground behaviour within an acceptable risk profile. The core value of the observational method is to use the *real* in-situ present ground conditions, as can only be encountered during construction. Its application will thus reduce the inherent ground uncertainty during construction, because the design assumptions of ground conditions and behaviour are proven and, where necessary, adapted by the observations of the actual ground conditions and behaviour.

There is, nevertheless, one ultimate prerequisite for the application of the observational method during construction: the design needs a certain degree of built-in flexibility to respond to the inherent differences in ground conditions that will be encountered during construction. If this design flexibility is not there, the observational method is simply not applicable (Peck, 1969). A close relationship between design and construction is thus required. For this reason I introduced the observational method in Chapter 10, as a *construction risk remediation* measure to be acknowledged already in the *design* phase. Design and construction are traditionally often different worlds about *predicting* and *providing* (Blockley and Godfrey, 2000). Modern *Design and Construct* (D&C) contracts are favourable in this respect, with design and construction being the responsibility of one sole party. That party may benefit from the observational method when reducing the sum of the design and construction costs, within the pre-set safety and quality standards, by a smart use of the observational method.

A characteristic disadvantage of the observational method was the relatively large and time-consuming workload, by interpreting the observations made and back-analysing the likely effects on the project. Today's modern and cost-effective *information and communications technology* (ITC) dramatically reduces this disadvantage. Large data sets of monitored ground behaviour are simply sent to databases by just a cell-phone, from many places all over the world. Nearly real-time interpretation and back-analysis of these data are possible with an increasing number of software packages. An SMS-message can act as an alert to a geotechnical expert, who just has to plug in his or her laptop computer to the Internet, to be able to check what could be wrong and how to act accordingly. This professional may very well be located on the other side of the world. In summary, today's ground-related monitoring data can easily be made accessible, anytime, anywhere and to anyone involved. Additionally, developments in modern *sensor technology* produce abundant sorts of multi-purpose sensors, some as small as a pin's head, which can be installed in the ground or just mixed within the concrete. Finally, *remote sensing* technology and satellite images have become ever more available at lower cost. Figure 12.3 summarizes these favourable developments and technologies for applying the observational method.

Because of all these developments, I expect a revival of the observational method, particularly within the GeoQ risk management framework. The combination of improved ICT-facilities, sensor technology and remote-sensing techniques may trigger some kind of digital revolution in the way we currently deal with ground-related design and construction. I predict a major shift towards much more flexible and integrated design and construction methods. With papers like those of Soudain (2000) I do not feel isolated with my statements. He presents cost-saving assessments between 5 and 25 per cent by a rigorous application of ICT in the construction industry. It is a challenge to realize these savings, and ground-related risk management by the observational method during construction may become one of the main drivers.

In short, the application of the observational methods within a risk management framework pursues cost-effective ground risk remediation, while chasing hidden opportunities as well. In particular, for the less serious ground-related

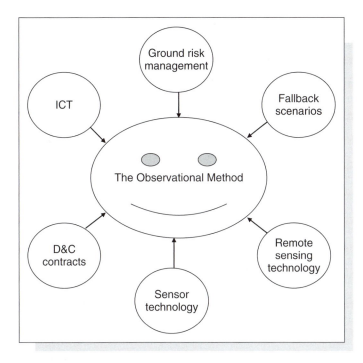

Figure 12.3: Developments and technologies for the revival of the observational method.

risks, reducing the risk *effect* can be attractive. Contrary to the concept of risk *cause* reduction, which is typically applied during design, risk effect reduction will only cost money when the risk indeed occurs, besides the costs for the (additional) monitoring and fall-back scenarios. Also, hidden opportunities of *better* ground behaviour than expected, which are revealed during construction, may be used. An example is the saving of 4900 tonnes of temporary supporting steelworks for a London based cut-and-cover highway tunnel, which represented 91 per cent of the original design (Glass and Powderham, 1994). Other promising savings are presented in the case studies at the end of this chapter. Apart from geotechnical and geohydrological aspects, the observational method also proved to be effective and efficient for the geoenvironmental type of projects (van Meurs et al., 2001). This will be further explored in Chapter 13 about the maintenance phase.

Additionally, the observational method is not restricted to risk management purposes only. Also *quality management* and quality assurance of a project, in particular on greyfield sites with a lot of interaction with adjacent structures, may demand rigorous monitoring during construction, as for instance presented by Savadis and Rackwitz (2004). They describe an extensive monitoring system during the redevelopment of the transportation infrastructure in the city of Berlin, after the union of East and West Germany. Berlin became Europe's biggest construction site in the mid-1990s. The impact of this infrastructure project on Berlin's urban life and environment could only be managed by the application of the observational method. This is where the concepts of risk management and quality management meet each other. In order to satisfy both the demands for risk management and quality management, the observational method can be extended to a *Building Risk Assessment* (BRA) method, as proposed by Burland (1995). First, the potential surface movements, as induced by excavations for instance, are analysed by this type of method. Then, the possible consequences of these construction activities on adjacent structures are classified in terms of the severity of expected damages (Portugal et al., 2005). In other words, a number of potential *damage scenarios* are identified and for each scenario a number of risk remediation measures or fall-back scenarios will be provided.

Finally, in (recent) history the observational method has occasionally been used as an instrument for *crisis* management, something went wrong in the ground and only the observational method could save the project. From now on we can give the observational method a positive boost in the direction of becoming a major ground *risk* management supporting approach.

Avoiding the pitfalls

By now we know about the many benefits of the observational method. However, as with most, if not all, methods in and outside construction, besides advantages there are inevitable limitations and pitfalls. The observational method is no exception and we have to acknowledge these to guarantee effective ground risk management. The following issues of concern were recognized by Peck (1969) and are still relevant:

1 *Fall-back scenarios* should be available for every unfavourable situation that may be disclosed by the observations during construction

2 The observations must be *reliable*, reveal the significant mechanisms and must be reported in such a way that immediate action is possible, when required

3 The mechanism of *progressive failure* must be recognized as being very difficult to observe before it actually takes place.

The first pitfall, an absence of a complete set of *fall-back scenarios*, has been previously explored in Chapter 10 about the design phase, because these scenarios need to be developed and planned during design. The GeoQ process facilitates the process of scenario analysis, including fall-back scenarios, by the structured identification and classification of ground-related risk. If sufficient attention is paid to ground risk management during all project phases, it must be feasible to avoid the first pitfall.

With regard to the second pitfall, the need for *reliable* observations, we again approach the soft and hard systems in construction, the interaction between the subjective people factor and the objective technical reality. A wealth of sophisticated technical systems can provide measured ground data of a high reliability and many sorts of monitoring equipment, such as earth pressure gauges, inclinometers and all sorts of devices to measure groundwater pressures, are available worldwide. However, certain mechanisms may be not *measured* but only *observed* by the human eye of an experienced geotechnical engineer or engineering geologist. The inspection of dikes in The Netherlands is an example. The detection of local small ground movements and cracks, which may trigger dike failure during high water conditions, is not yet sufficiently reliable detected with only technical systems. These dikes are also visually inspected by professionals on a regular basis and with an increase in intensity during expected high water. However, human observations and the resulting interpretations embed an inherent subjectivity in the conclusions that will be drawn from the ground data observed. I recall the differences in (risk) perception and often we only see what we are used to seeing, as demonstrated by a few of my overseas experiences in Box 12.1.

Box 12.1 demonstrates the need for reliable observations and just two brief examples from practice demonstrate the difficulties. This teaches the value of a thorough preparation with regard to the specific regional site conditions, before entering the construction site.

However, this may still be insufficient. The pitfall of the unaware and biased individual professional can be even better avoided by appointing a (small) team of two or more professionals for providing the observation and interpretation, with the team members preferably having different experiences and backgrounds. In addition, these teams of professionals do also increase the probability that all most important (failure) mechanisms will be recognized. A professional project organization, with clear responsibilities, should furthermore guarantee appropriate actions being taken immediately, if freshly observed monitoring data reveal unexpected deviations.

Finally, it should be realized that so-called *progressive failure* may be missed by monitoring and observation (Peck 1969). For instance, stiff clay layers in a cohesive soil or relatively hard layers in a tropical residual soil may develop high resistance at small strain. If their limit strength is reached, these layers suddenly fail and transfer their load to other less resistant layers. This can trigger a number of subsequent failures, which may result in a slope failure. Unfortunately, such failure starts *before* actual movements can be observed. Therefore, if the risk of progressive failure is identified, which typically demands a detailed risk analysis during design, appropriate risk cause or effect remediation measures should be taken, instead of relying on an, in this respect, inadequate observational method.

Appropriate *monitoring programmes* further avoid the pitfalls of the observational method as a risk remediation tool during construction. As mentioned, in today's modern world the observational method will be largely facilitated by all kinds of technical monitoring systems and programmes. As with any programmes, monitoring programmes should be driven by clear objectives. By the application of the observational method, within the open framework of GeoQ ground risk management, the ultimate goal of monitoring is to facilitate controlling ground-related risk. As a major benefit the monitoring programme becomes risk driven,

---

Box 12.1   The art of reliable observations on sites in the desert and the tropics

My profession as an engineering geologist brought me to several places in the world, with entirely different ground conditions. My first overseas project site was located in the Middle East. It was a large construction site, where an airbase was under construction. I was responsible for the engineering geological site characterization. One of the main challenges was to determine the required excavation level of a number of large construction pits. Most of the typical desert site conditions were available, including the so-called *collapsible* ground. An experienced engineering geologist, Ard, being my supervisor and visiting the site on a regular basis, made me aware of this fact. This type of ground has soft rock properties when in dry conditions, which is obviously the common situation in a desert. However, when this ground material becomes saturated with water, it transforms quickly to a soft soil, which results in an easy collapse of the soil, when loaded.

This collapsible ground behaves fuzzy from a contractual point of view. At first observation it behaves like rock, allowing the subcontractor to be paid for rock excavation. A more thorough observation of this material, when wetted, raised the conclusion that there are also arguments to pay the excavation of this material on the basis of the lower soil excavation rates. In particular, if the subcontractor applies water during his excavation activities. It is a typical source of a dispute during construction.

One of my next missions brought me to Malaysia, where I encountered typical tropical ground conditions. Severe and particularly chemical weathering of rock masses is an occasional phenomenon that is typically encountered in these areas. Drilled samples revealed stiff clay and, coming from The Netherlands, I was familiar with stiff clay, caused by over-consolidation during the Pleistocene ice ages. Only after careful observation of the samples, which revealed also the presence of small rock fragments, I realized that this stiff clay was the end product of serious weathering. I remembered the highly variable weathering profiles, which implies the risk of large variations in sound bedrock levels at relatively short distances. Furthermore, boulder-like elements of less severely weathered rock might cause problems during pile driving. These are some examples of typical triggers for ground-related risk that need to be identified during a site investigation, by careful observation.

---

which might convince sceptical clients or contractors about its necessity and value for money. By linking monitoring to risk, it is possible to define programmes with the optimum benefit-to-cost ratio, in view of the risk tolerance of the parties involved. In other words, monitoring can be tailor-made to risk remediation.

Therefore, any monitoring programme should be based on the answers to two questions (van Staveren, 2003):

1 What to monitor?

2 How to monitor?

The first question is a strategic one: *what* to monitor depends on which risks to control during construction. The identified and classified ground-related risks, as a result of the previous GeoQ steps, should serve as a basis to answer this question. Ground-related risk to be controlled depends obviously largely, if not totally, on the contractual risk allocation, as can be agreed in the Geotechnical Baseline Report. With regard to the *initial ground assumptions* on which the monitoring programme will be based, several approaches are possible:

1 The most likely ground conditions (dream scenario)

2 The worst case ground conditions (doom scenario)

3 Several likely ground conditions (several scenarios)

4 A probabilistic approach of several ground condition scenarios.

The first option is obviously over-optimistic and can better be omitted. The second option is much safer, however, perhaps too negative. The third choice is probably more expensive, because more than one scenario requires probably more instrumentation and observation. However, monitoring focused on a few scenarios will increase the number of opportunities that can be caught as well, such as stronger ground than expected. The last choice, the probabilistic approach, will involve some more investment in the preparation of the monitoring programme. This can be an excellent investment for the rather complicated and large projects, because the costs of the monitoring programme can be balanced with the probable benefits of it. What type of approach to adopt depends entirely on the type, complexity and accepted risk profile of the construction project.

The second question about the monitoring programme is of a more operational or tactical nature. Once it is clear *what* needs to be monitored, in terms of ground-related risk and associated ground behaviour, the question about *how* to monitor most efficiently demands an answer. In this respect it is good to remember the words of Dunniclif and Green (1988): 'Every instrument on a project should be selected and placed to assist with answering a specific question: if there is no question, there should be no instrumentation.'

The required *accuracy, reliability and usability* of the monitoring programme, in view of the expected ground conditions, need to balance the unavoidable budget and time constraints. It is considered beyond the scope of this book to answer this important but also rather complicated question about how to monitor and additional specialist advice will often be available. Books like the one of Dunniclif and Green (1988) provide a lot of information about all sorts of ground monitoring equipment. Dunniclif and Powderham (2001) stress the importance of considering monitoring and its geotechnical instrumentation as a professional service, rather than a lowest price construction necessity. They highlight as well the importance of a *team* environment to ensure best value for money.

To conclude, we explored the observational method in combination with monitoring as an effective tool for remediating risk effects, which supports the GeoQ ground risk management process during construction. The inherent pitfalls of the method can be largely avoided with due care. However, due to the inherent ground uncertainty, no doubt a certain degree of residual risk remains. Unfavourable differing ground conditions may still surface during construction. How can we deal effectively with the required ground dispute resolutions, in these cases?

Differing ground conditions during construction: what we can do

Construction proves the reliability of the forecasts about ground conditions. Differing site or ground conditions typically arise in the construction phase, when the actual ground conditions and behaviour are encountered during activities such as excavation, boring, and injecting. In other words, differing ground conditions are an inherent part of construction. Consequently, also differing ground condition claims are normal occurrences in the construction phase (Poulos, 1998). Therefore, in this section I will not answer the question how to minimize the probability of a differing ground condition claim, because the remaining part of the book concerns this item. In this section I will introduce some ways for dealing with a differing ground conditions situation during construction, within the framework of the GeoQ ground risk management process. As the reader would expect, this section builds further on Chapter 11, where the concept of the Geotechnical Baseline Report (GBR), together with the Differing Site Conditions (DSC) clause and Dispute Review Board (DRB) were introduced, discussed and illustrated by a number of cases from several countries.

First, I briefly discuss the situation where a differing ground condition occurs and is solved in mutual agreement, without a dispute, by the joint application of the DSC, GBR and DRB, which reflects the ideal project situation. Second, I discuss those common situations in which a dispute does arise about a differing ground conditions claim, either because the construction project does not apply the combination of a DSC, GBR and DRB, or because these supporting acts do not result in an acceptable agreement between the parties involved.

Differing ground conditions in agreement

Even if the GeoQ process has been applied throughout the entire construction project, differing site or ground conditions are to be expected during construction. That should not necessarily be considered as a risk management failure, because the application of the GeoQ process will have prepared us well for these differing conditions. A DSC will be part of the construction contract, which means that agreed differing site conditions are to be compensated by the client. The GBR will be of great help in deciding whether the ground conditions are indeed differing or not. The contractual baselines, as provided by the GBR, will act as a basis to appraise and judge any differing ground conditions. Normally, the contractor has to prove the actual ground conditions encountered are different, with regard to the agreed baselines, for which the contractor has to apply similar ground investigation techniques as has been used for establishing the baselines in the GBR.

For example, during an excavation a sandstone rock tends to be strong, with unconfined compressive strength (UCS) values over 50 MPa, while the baseline for the UCS was set at 30 MPa, representing moderately weak to moderately strong rock. In addition, the discontinuity pattern of the rock mass is less favourable to excavation. Therefore the excavation activities are much more difficult as well as slower than scheduled. Based on the GBR, it is fairly straightforward to judge this situation: a differing ground condition is there, based on the agreed baselines in the GBR. The contractor will be compensated, either by a pre-set financial arrangement or by a reasonable change order.

In a lot of cases the representatives of the client and contractor can agree on differing ground conditions just by comparing the additional ground investigation data with the baselines of the GBR. If not, then the DRB has to decide whether the ground conditions are differing in an unfavourable way from the baselines or not. The application of the Geotechnical Baseline Report should also prove its added value in this situation, because it helps to make explicit and measurable to the DRB whether ground conditions actually differ or not.

By following this procedure, as an integral part of the GeoQ ground risk management process, the majority of situations with differing ground conditions should be arranged between the parties in a satisfying way. Consequently, a lot of common confusion, as well as disputes about the contractual consequences of unforeseen ground conditions, may be reduced or avoided, although I recognize that the GBR concept is still subject to a lot of discussion in our industry, as earlier disclosed. The prevailing reality is often not so favourable to us.

Ground dispute resolution

In our ideal project, the combination of the DSC, GBR and, when required, the DRB should be adequate to agree about any differing site or ground conditions in a cost-effective and efficient way. Both parties should be confident with the agreement and continue the project on pleasantly speaking terms. How different reality often turns out to be. A ground *dispute* arises when the contractor encounters a differing ground condition, either a Type I or Type II situation, which is not agreed by the client. Normally, a contractor will only raise a differing ground condition when that condition has an unfavourable effect on his activities. I recall the previous rock excavation example. The contractor may have to mobilize another more suitable excavation equipment at additional cost. Also a delay may occur, because of the harder excavation activities. Most contractors will claim compensation for additional costs from the client in order to return a reasonable profit from the project. The contractor may also ask for some relaxation of the time schedule. If the client does not agree, a ground-related dispute has been born. Obviously, the probability of this situation is much higher when there have been no GBR and DRB established for the project.

As an ultimate consequence, both parties see each other in court. On the other side of the dispute continuum, just a single meeting between representatives of both parties dissolves the entire dispute, resulting in a mutual acceptable solution for both parties.

The good news is a rising awareness, as well as willingness, to solve ground-related disputes with less rigorous means than bringing the case into court. This development started some years ago. In 1988, the Association of Engineering Firms Practising in the Geosciences (ASFE) in the USA published a manual which included a whole spectrum of *Alternative Dispute Resolution* (ADR) techniques (Caspe, 1998). The manual starts with negotiation and moves via mediation, arbitration, mini-trials and private litigation towards official litigation, just to mention the most significant techniques (Association of Engineering Firms Practising in the Geosciences, 1988). Obviously, combinations of these methods are also possible. In addition, the ASFE manual suggests resolution through experts, which seems comparable with the DRB. In the UK there is also an awareness that litigation and arbitration, as well as mediation, in order to solve disputes on ground-related matters are often not satisfactory (Turner and Turner, 1999). Therefore, in view of modern ground dispute resolution within the GeoQ process during construction, I have selected two alternative dispute resolution techniques for more detailed consideration: *mediation* and *arbitration*. I compare these techniques with the ultimate *litigation* solution in court. Figure 12.4 presents the three ground dispute resolution options with their main characteristics.

In the following concise discussion of these ground dispute resolution options, the presence of a Differing Site Conditions clause, as presented and discussed in Chapter 11, has been assumed to be a pre-condition. If a DSC is not applicable, it is likely that the *common law* rules. This implies it is the single contractor's responsibility for dealing with differing ground conditions, independent of its seriousness and effects on the project. Therefore, in my strict interpretation of the common law, a dispute about differing site conditions is by definition no issue in this situation.

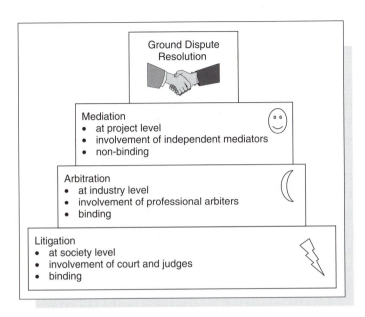

Figure 12.4: Three options supporting ground dispute resolution.

The three ground dispute resolutions options of mediation, arbitration and litigation to be discussed assume a DRB is either not available, or not able to solve the dispute to the full satisfaction of the project parties. How can the application of the GeoQ process during construction, preferably together with the GBR and all ground-related risk knowledge of the preceding project phases, assist these three ground dispute resolution options?

Let us start with the option of *mediation*, which is defined here to be a *non-binding* and confidential process. A neutral mediator assists the parties in reaching an agreement about how to solve the dispute (Caspe, 1998). The mediator should understand the interests as well as the personalities of the individual representatives involved in the dispute. The latter typically reflects the soft systems approach, with due attention to the people factor. The results of the GeoQ process, as available when the mediation takes place, may largely facilitate the mediation process. As a result of the GeoQ process, clear ground-related risk registers, including risk remediation measures, are available and can be used. Perhaps even a GBR is present. All this information may help to create more insight into the effects of the differing ground conditions to the contractor and the client, in view of their likely different risk perceptions. This may serve as a basis for mutual understanding and result in an acceptable agreement for both parties. In that case the mediation has succeeded.

If the mediation does not give an acceptable dispute resolution, the next step may be so-called *arbitration*. Contrary to mediation, arbitration is a binding and formal process, following a specific set of procedures, such as for instance prepared by the Construction Industry Arbitration Rules of the American Arbitration Association (Caspe, 1998). The involved arbiter should be an acknowledged expert in the field of ground-related disputes. However, unlike a trial, the strict rules of evidence do not apply. Obviously, as in the case of mediation, also during arbitration, results of the GeoQ process may significantly help to arrive at a fair and acceptable solution. The end-result of an arbitration is generally not subject to appeal. However, if there are extraordinary circumstances, so that the outcome is subject to appeal, the next and final step is to start a process of *litigation*. Obviously, the decision to litigate can also be made directly, without the intermediate steps of mediation and arbitration.

The impact of litigation on all parties involved should not be underestimated. They can be major and should raise serious awareness and concern, preferably well before starting any litigation process preparations. Besides the cost involved to hire the necessary lawyers and technical experts, as well as for research and testimony, considerable time should also be reserved. For complex situations, procedures of several years are no exception. In addition, the relationships of the project parties are often highly disturbed and may even result in a 'pathological situation' (Caspe, 1998). Contrary to arbiters, judges and juries are seldom experts of differing ground conditions and construction. Therefore, both parties have to mobilize their (hired) experts. These are likely to be biased to the perspectives of their respective clients, which does not make life easy for the judges and juries involved. Moreover, most countries operate largely different court systems, which can raise additional complexity if an international consortium starts litigation for an overseas construction project with an overseas client. As a result of the presented drawbacks, for most cases it is expected to be much more constructive to avoid litigation. However, if litigation becomes really unavoidable, it is hoped that the data from the GeoQ process may have a positive effect on the duration and outcome of the litigation process. Box 12.2 provides an example of a consideration of bringing a case into court or not.

It is clearly up to the reader to judge the most appropriate ground dispute resolution method for the project. General rules for their application would create a false certainty, except the only one I have: *it depends*. Obviously, it can also be very effective to use alternatives or combinations of the alternative resolution methods, as presented for instance by the ASFE. What I would recommend, however, is at least to consider the sequence of mediation, arbitration and litigation. Simply because costs, time, energy and also frustration with all parties involved are expected to increase in the same order. In other words, a lot of these precious resources may be saved by reaching an acceptable ground dispute solution, without meeting at court and may be even without arbitration and mediation.

Nevertheless, unforeseen ground conditions remain a fact of construction life. Our challenge is thus twofold by avoidance and curing: to minimize occurrence of any differing ground conditions by operating a thorough ground risk management process and to settle them effectively and efficiently, once they are there. The application of the GeoQ process may serve as a catalyst to meet this double challenge. In this respect, Martin (1987) and Gould (1995) suggest the observational

---

Box 12.2 Going to court or not: that is the question

The first bored tunnel in The Netherlands, the Westerschelde Tunnel, has a length of 6.6 km. The tunnel consists of two tubes, each more than 11 m in diameter. Its deepest point is at some 60 m below sea level and the tunnel crosses geologically highly complex ground conditions, varying from rather soft and loose deposits to very hard and over-consolidated so-called Boom clay.

When the tunnel boring machines reached their deepest point, the shields of both became distorted in the highly over-consolidated and stiff clay layers and the boring process had to be stopped. A typical situation of unforeseen ground behaviour.

The Design, Construct and Maintenance contract (DCM) included a penalty for delay of 136 000 euros per day, for a maximum of 100 days. However, the serious boring problems took some 8 months in total to be solved. Consequently, the 100 days of penalty were exceeded and the contractor's consortium lost a fee of 13.6 million euros. The client seriously feared that the contractor would continue the project rather cost- than time-driven, to minimize the unfavourable financial consequences of the lost fee, by cost-savings of cutting overtime and other measures increasing the total contraction time of the project. This situation was highly worrying for the client, who would loss toll income with a later project completion. Therefore, the client had to decide whether to go to court or not, to force the contractor to speed up the project.

The client decided to negotiate first, before going to court. It resulted in a so-called 'package deal' with the contractors, in which the completion date has been shifted and the penalty would be waived. In addition, a bonus system was agreed, in which the contractors earned 68 000 euros for every day of completion before the agreed date. This positive incentive worked quite well, the tunnel was completed 8 months before the newly agreed date. This fast completion was particularly the result of optimized logistics and organizational improvements (Heijboer et al., 2004). In the mean time, no further problematic differing ground conditions occurred. The final costs were only slightly higher than budgeted. Apparently, it appeared to be a good decision to start with negotiations instead of going straight to court in order to resolve the problems resulting from seriously differing ground conditions.

---

method to support disputes, which has been used since its first time in 1984 in Washington with marked success on several rock tunnels with complex ground conditions in the USA. Figure 12.5 demonstrates the ultimate GeoQ objective within the dispute resolution process: spending of less costs and less time to solve the dispute, for all parties involved.

Munfah et al. (2004) provide a number of increasingly used construction management policies for delivering additional contributions to the differing ground conditions challenge. These include prequalification of contractors, full disclosure of all available ground-related information, the inclusion of partnering, the stimulation of value engineering, and last but not least, the provision of an owner controlled insurance programme.

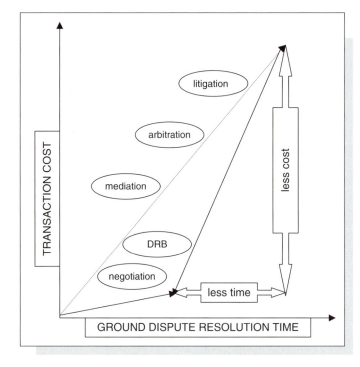

Figure 12.5: The GeoQ objective in ground dispute resolution.

Case studies

The purpose of the two cases to be presented in the following is to demonstrate a few ways for applying the GeoQ process in the construction phase of a project. In the first case we revisit the Betuweroute railway project of Chapter 10. Embankments, with a perceived severe risk of slope instability and failure, needed to be constructed within a minimum of time on very soft soil. The application of the observational method in combination with a rigorous on-line monitoring programme proved to be highly successful. As will be described, a similar approach proved to be successful in Australia. The second case describes how the GeoQ approach supported the recovery of a project in crisis during construction.

Stability risk control by the observational method

This is where we recall the Betuweroute railway project, the second case study of Chapter 10. This Design, Construct and Maintenance (DCM) project with an alliance or partnering contract has entered the construction phase. We consider again the section between the cities of Sliedrecht and Gorinchem. This most complicated part of the new railway passes through low-lying areas with very unfavourable ground conditions with very soft peat and organic clays. The thickness of the soft layers is approximately between 10 and 12 m. Some typical figures for the ground experts among the readers, volumetric weights range between 10.5 and 14kN/m^3 and the undrained shear strength of these layers varies between 5 and 15kN/m^2 (Molendijk, Van and Dykstra, 2003).

With regard to the first GeoQ step of *gathering information*, we classified this section already as a typical greyfield project, while running parallel to sensitive existing infrastructure, the existing railway and the busy motorway A15. Consequently, in addition to the bad ground conditions, minimizing the disturbance of many thousands of daily users of the railway and motorway provided an extra challenge for the construction team of the alliance.

In view of the second and third GeoQ steps of ground *risk identification* and *risk classification* at the start of construction, the team encountered, among others, a serious soft soil-related embankment stability risk, particularly in view of adverse effects on the adjacent infrastructure. The team had already decided to go beyond the conventional design standards during design because reducing the stability risks simply by a conservative design would cost several million euros extra, in advance. The ground professionals in the consortium trusted their combination of experience and engineering judgement to reduce the stability risk to an acceptable level during the construction phase.

All identified and classified ground-related risks were allocated and presented in a Geotechnical Baseline Report. The consortium shared the responsibility for the majority of ground-related risk and a risk contingency budget has been reserved, in which all consortium partners got a share. Any financial consequences of risk remediation would be paid from the contingency budget, while any cost savings by optimizations would flow back to it. At the end of the project, the remaining contingency budget would be shared between the consortium of the client and the contractors. The entire construction team was therefore highly motivated to ground risk management, with a mutual interest for a maximum of risk reduction and construction optimizations, at a minimum cost. This willingness resulted in innovative solutions during construction.

This case will now focus on a 9 km long embankment section, where the Betuweroute runs just a few metres from the existing railway. Two ground related-risks, classified as serious, were of particular importance:

1 Slope instability and collapse of the new embankment adjacent to the existing railway, caused by the poor strength of the underlying soft ground layers

2 Unacceptable horizontal and vertical deformations of the existing infrastructure, as a result of the loads of the embankment of the new railway.

For the embankment slope instability a *risk cause reduction* measure was selected during design. The embankment would be built up rather slowly, in layers of about 1 m thick. Then a period without construction activities was planned, in which the groundwater overpressures in the underlying soft layers caused by the embankment could dissipate. If a height of 2.5 m was constructed at once, instability and failure of the embankment was expected. However, additional calculations with strength parameters from a regional data set, based on routine laboratory investigations, showed that even the first sand deposit of 1 m could already lead to embankment failure. In other words, the risk cause reduction according to the design appeared to be not appropriate during construction.

For the second risk of unacceptable deformations of the existing railway, a 9 km long sheet pile wall has been designed in the most critical area in between the existing and new embankment. This measure can be considered as a *risk effect reduction*.

After this fourth GeoQ step of *risk remediation*, the next step of *risk evaluation* followed. In view of their actual situation, the geotechnical experts of the consortium's team considered the ground characterization as not realistic and too pessimistic. To prove their engineering judgement, they decided to apply the observational method in combination with a rigorous monitoring programme in two stages. First, a test embankment was constructed and carefully monitored during its construction. The embankment was constructed by 0.5 m layers of sand. Horizontal and vertical deformations, as well as changes in groundwater pressures, were measured by porewater pressure meters, settlement plates and inclinometers.

The results were considered as very promising and confirmed the engineering judgement of the team. In fact, failure of the embankment occurred after only a height of 5 m was reached. The ground was locally considerably *stronger* than had previously been assumed during design. Also the horizontal and vertical deformations near the embankment were much smaller than initially expected. Therefore, it was decided two perform two optimizations during construction and the results of the test embankment served as a basis for adapted strength and stiffness parameters. First, the embankment could be constructed much faster by the revised approach. Second, and in addition, the team decided to omit the planned protective sheet pile wall between the new and existing embankment. As an additional risk remediation measure, to monitor the risks of these modifications, the observational method continued during the entire embankment construction by the operation of a rigorous on-line monitoring system.

A large number of instruments were installed to measure changes in porewater pressures, as well as horizontal and vertical deformations. At geologically relevant cross-sections, where a large degree of geological heterogeneity could be expected, advanced so-called vibration wire piezometers were used. The 'on-line' availability of all of the monitoring results proved to be a very successful aspect of this detailed and advanced ground characterization by monitoring. All monitored data were immediately sent to a database, which was accessible via the Internet for all authorized professionals of the parties (Molendijk and van den Berg, 2003).

The database also served as a basis for the final GeoQ step of *mobilizing*, in which all relevant ground-related risk data needed to be prepared for the next phase of maintenance, which was part of the consortium's DCM contract. Figure 12.6 shows the concept of on-line monitoring, including the storage of all data in the database.

The presented risk-driven observational method supported extensive monitoring, deleted the initial requirement of 9 km of sheet pile, which saved the alliance some 4 million euros. The additional monitoring cost was approximately 0.2 million euros, which gives a cost:benefit ratio of 1:20 for this design optimization. The net savings of 3.8 million euros were shared by the consortium partners, according the alliance contract (Molendijk and Aantjes, 2003). It is remarkable that the contractors earned money by *not* installing a sheet pile wall. In addition, the much faster construction of the embankment, without standing time, provided a

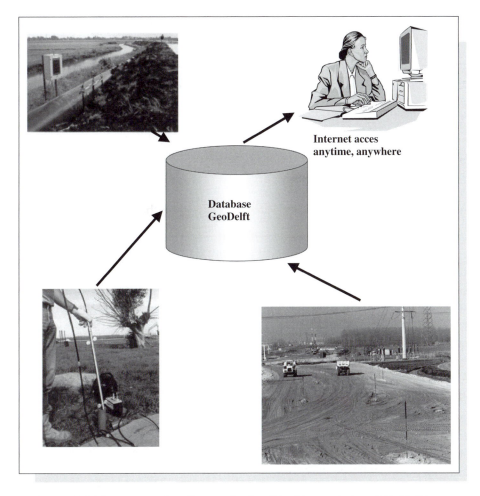

Figure 12.6: The concept of on-line monitoring (© with permission of GeoDelft).

welcome effect to the project's schedule. Any potential dispute about differing site conditions was negotiated and agreed on site, by the directly involved represent-atives of the partnership. The Dispute Review Board did not need to undertake any serious action. The entire Sliedrecht–Gorinchem part of the Betuwe railway project was completed well within planning, while the total project costs savings summed up to some 25 million euros, which is approximately 10 per cent of the original project budget.

Ameratunga et al. (2005) present a similar GeoQ-like and risk-driven approach on the other side of the world, for a seawall construction project in Brisbane, Australia. Box 12.3 presents a number of remarkable similarities of the Australian and Dutch approach.

Box 12.3 supports the viability of a GeoQ-like ground risk management approach for a nearshore construction project. Like Australia and The Netherlands, there are many other countries and regions facing similar challenging ground conditions and project constraints in which a GeoQ approach during construction is expected to add value in terms of ground risk and opportunity management.

A tunnel in geohydrological crisis conditions

In view of the second case study, I would like to start with some other words of Peck (1969):

> Whenever construction has already started and some unexpected development has occurred, or whenever a failure or accident threatens or has already taken place, an observational procedure may offer the only satisfactory way out of the difficulties.

How true these words appeared for a design and construct tunnel project in the political capital of The Netherlands, the city of The Hague. During construction of this cut-and-cover tunnel, on a typical greyfield site in the middle of the main shopping area, a number of serious and acute groundwater problems forced the entire construction process to stop. In total there were five subsequent and serious water leakage problems, mainly resulting from a failing grouted sand layer. This project, named Tramtunnel, became known by the public as the Tramtanic.

Eventually, measures to remediate the surplus of groundwater appeared remarkably to result in an unacceptably low groundwater flow, which created a new and major risk. This latter issue is the subject of this case, in which I present a few highlights to demonstrate the possible role of the GeoQ process in a crisis situation. At that time, the GeoQ process could not have been applied as explicitly as presented in this book, due to its early development stage in the early 2000s.

What type of crisis are we talking about? During construction of one of the tunnel's deep stations, the contractor's consortium faced a problem. The drainage system in the building pit was not working well. This resulted in unacceptably high groundwater pressures in the strata underlying the excavation. Without

---

Box 12.3 An Australian approach to stability risk management

The Moreton Bay project involves design and construction of a 4.6 km long sea-wall near the port of Brisbane. The project applied a risk-driven observational method with an extensive monitoring programme. The following striking similarities of this project with the Betuwe railway project I retrieved from the paper by Ameratunga et al. (2005):

- Operating a design and construct *alliance* between the client and team of consultants and contractors

- Facing considerable project *constraints*, for instance by the close proximity of the Moreton Bay Marine Park

- Presence of *very soft soil conditions*, with undrained shear strengths as low as 3 to 5 kPa, and clay layers extending over 30 m below the seabed

- Soft soil bearing capacity and *embankment* stability as a *critical* issue

- Serious *time pressure* which did not allow a staged construction of the embankment

- Applying rather *low factors of safety* during construction, typically between 1.15 and 1.25, despite high consequences of failure

- Using *instrumented trial embankments* to obtain better understanding of the feasibility of the proposed construction methods

- Performing *back-analysis* of calculations with the monitoring results of the trial embankments

- Using a Geotechnical Work Method Statement (GWMS) *jointly* prepared by the design and construction teams to manage risk during construction

- Operating an extensive *monitoring programme* during construction for monitoring ground and construction behaviour, of which the results were stored in a database

- Allowing *flexibility* during construction to apply any remedial measures when required.

---

relief, the pit-bottom would burst open, which would result in large deformations of the diaphragm walls and subsequent intolerable damage to the adjacent multi-storey-buildings that are part of the shopping centre of The Hague. Consequently, the excavation process and construction were put on hold.

Also in this type of a rather critical situation, the first GeoQ step would have been that of *gathering information*. A lot of ground data were available during construction and *risk identification* and *risk classification* would confirm the collapse risk of the deep station bottom as very serious, acute and to be avoided at any cost. However, the problem became even more serious in view of the next GeoQ step of selecting *risk remediation* measures. Neither fall-back scenarios, nor appropriate risk cause or risk effect measures were available. Even worse, the risk cause was not at all clear: *why* did the groundwater drainage system not work as expected and required?

In order to get out of this highly unfavourable situation for all parties involved – a construction on hold, without a solution how to proceed – a number of external experts were consulted, including the Delft University of Technology and GeoDelft. A careful risk-cause analysis was performed and Box 12.4 presents a few more details, for those readers interested in the geohydrological and geochemical details.

Box 12.4   Some details of a geohydrological risk-cause analysis

A detailed risk-cause analysis of the malfunctioning of the groundwater drainage system related clogging of the extraction drains to a gel layer, which had been injected into sand layers underlying the excavated deep station's bottom. The purpose of the gel was to make these sand layers impermeable, to prevent excessive groundwater inflow in the tunnel during construction. Well, that objective clearly has been reached, however, together with an unforeseen side-effect: impermeable drains to relieve excessive groundwater pressures in the underlying layers. As became clear by a geochemical analysis, the hardening process of the gel layer released caustic soda into the soil. A rather high caustic soda concentration would dissolve the locally present organic material of peat lenses in sand layers. A groundwater flow removed this material and, at a certain distance from the gel layer, the organic material flocculated. This caused the clogging of the ground surrounding the drains. Extensive laboratory tests were carried out to test the validity of this hypothesis and these experiments confirmed the theory. At least the main risk cause appeared to be known (Luger et al., 2003).

This knowledge about the risk cause provided an opening towards an effective risk remediation measure: to re-operate the dewatering system. Installation of sand piles with a valve system was selected as the most appropriate risk remediation measure. The observational method, by operating a rigorous monitoring programme, was the only possible way to continue the deep station excavation and to complete construction within an acceptable residual risk profile. The on-line monitoring system enabled continuous information about any changes in the groundwater pressures, as a way of *risk evaluation*, representing GeoQ step five. The monitoring data were stored in a database and could be *mobilized* when required (GeoQ step six) during the operation and maintenance phase of the tunnel. The newly installed groundwater drainage could be regulated, as well as the excavation process of the deep station, when necessary. By using this approach, it was possible to continue the excavation safely, about two years later, and to finish construction without facing similar problems.

The observational method, in combination with on-line monitoring, was essential to get out of the crisis and to complete this rather unfortunate construction project. Additional project costs have risen to over some 50 per cent of the original contract price and the construction time has been doubled. This second case demonstrates the actual validity of Peck's statement of 1969.

Summary

The proof of the pudding remains in the eating. This chapter has demonstrated the application of the GeoQ process in the construction phase. Two main approaches, contributing to the fourth and fifth GeoQ process steps of *risk remediation* and *risk evaluation*, are presented and discussed.

The first method revisits the observational method, from a risk management perspective during construction. This method is appraised as a promising risk remediation tool during construction within the GeoQ process, in particular to reduce the risk effects. In combination with today's modern design and construct contracts, widely available sensor technology, remote-sensing techniques and modern ICT-facilities, the application of risk management will be highly facilitated by adopting the observational method in an early stage of the project. Three main pitfalls of the observational method are the absence of fall-back scenarios, a lack of reliable observations, as a result of biased perceptions and inexperience with particular regional site conditions, and the mechanism of *progressive* failure. The latter cannot be effectively observed before it actually takes place. Apart from due attention to these pitfalls, their bypassing will be provided by the design and installation of an appropriate monitoring programme.

The second method of ground risk remediation during construction covers the role of the GeoQ process, including the concept of the Geotechnical Baseline Report (GBR), when differing site or ground conditions occur. These differing conditions will remain happening, even while applying the entire GeoQ process, as a result of inherent ground uncertainty. Therefore, in addition to applying the GeoQ process, it is better to be prepared with a number of ground dispute resolution options as well.

Finally, two case studies are presented, in which the GeoQ process played a dominant role. In the first case we revisited the Dutch railway project with a severe risk of embankment slope instability, which threatened adjacent infrastructure. An Australian nearshore seawall project reveals striking similarities, with regard to the project challenges and applied risk-driven approach. The second case demonstrated how the structured approach of the GeoQ process may facilitate a project in crisis during construction. Contrary to the first case, which concerned mainly *geotechnical* risks, the latter faced major risks of *geohydrological* origin. For this case applying the observational method in combination with on-line monitoring proved to be the only effective and efficient way out of the crisis.

13 GeoQ in the maintenance phase

Introduction

After probably years of feasibility studies, pre-design, design, contracting and construction we arrive finally in the very last phase of the GeoQ ground risk management process: maintenance during the project's operational lifetime. Finally, we are there: the precious construction project has been realized. Likely, we battled a lot of ground-related risks during the preceding project phases. The quality of design has been tested during construction, as well as the type, quantity and quality of probably more than one ground investigation. Perhaps we encountered differing site conditions that had to be resolved, by negotiation, applying a Dispute Review Board, mediation, arbitration or even litigation. In the latter case we even may still be involved in a court case.

Anyhow, to build forward on the examples in the introduction of the preceding chapters: the bored tunnel has been completed and is ready for operation over the years to come. Ideally, the result of the preceding construction phase is a well-completed construction project, within budget, planning, and risk profile, and according to the pre-set safety and quality standards. Figure 13.1 presents the maintenance phase in the entire GeoQ process.

By the application of the GeoQ process in the earlier project phases we reduced the number of ground-related surprises for the current maintenance phase to a minimum. Nevertheless, however rigorously risk-driven and adaptive the project has been, some uncertainty will remain to accompany us during the project's operation. We are still dealing with ground and its inherent uncertainty and some ground risk remediation measures have to prove they are fit-for-purpose during operation. A typical example is the risk of unacceptable settlements for infrastructure projects, which may involve safety concerns for its end-users.

In this last phase, the GeoQ ground risk management process is applied for risk-driven and cost-effective maintenance during the operational life time of the project, within the project's operational specifications and the risk tolerance of those parties still involved. Preferably, also in this final phase, all of the six GeoQ steps are taken to identify, classify and manage any changes in the residual project risks.

We will revisit a few ground risk management methods that are of particular importance in the maintenance phase and a new method is introduced as well. We will start with

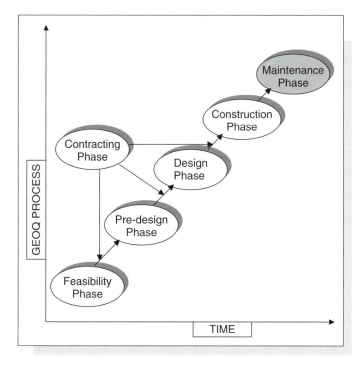

Figure 13.1: The maintenance phase within the six GeoQ phases.

the life cycle concept, as introduced already in Chapter 8, with a focus on cost-effective and fit-for-purpose maintenance. Then a new and risk-driven approach for appraising the maintenance and reinforcement needs of existing dikes will be introduced. These issues are a major safety concern in many areas in the world.

Next, we will revisit the baseline approach for risk allocation, as presented in Chapter 11, and we will concentrate on its role in dispute resolution about differing site conditions after project completion. These differing conditions may have an adverse effect on the project's anticipated maintenance programme. Obviously, monitoring remains important to obtain information on (different) ground behaviour in these situations. These three ground risk management tools can be related to the fifth GeoQ step of *risk evaluation*. Finally, attention is given to the sixth GeoQ step of *risk mobilization*, by a concise introduction of the main forms of information and communications technology (ICT) that can be used for this last step. After the discussion of these four ground risk management methods, two case studies, with the GeoQ process in a supporting role, are presented. As usual, the chapter finishes with a summary.

Ground risk management during operation and maintenance

Life cycle concepts for cost-effective maintenance

A large part of the construction industry attention goes to constructing and build-ing *new* projects. However, the end of construction is the start of the project's operational life and sometimes the enormous costs involved with the operation and maintenance of *existing* projects seems to be a bit undervalued. With regard to domestic housing in The Netherlands, the yearly costs of maintenance and renova-tion are more or less the same as the total yearly budget spent on new construction projects. In the infrastructure segment, the total budget spent on maintenance and renovation is even some 70 per cent of the total budget spent on new projects (van Staveren, 2001b). In addition, we should realize that the end-users of the constructed object add far greater value to it than the value represented by the total construction costs (Blockley and Godfrey, 2000). Let us consider the use of a bridge. Many thousands of people may cross that bridge every day during its entire life time of 50 or perhaps even 100 years. Imagine the value of that activity, by summing up the hourly rates of those people during these 50 or 100 years.

Two conclusions can be drawn. First, given the enormous cost of maintenance and operation, it would probably pay to reduce this cost by maintenance-effective design and construction. Second, given the enormous value that is represented by the end-users of construction projects, the life cycle *value* should not only be dominated by its *costs*, but also by its *purpose*. Blockley and Godfrey (2000) suggest expanding the life cycle *costs* approach to some form of life cycle *purpose* approach. This may require a detailed analysis of the future use and users of projects, in which the concept of scenario analysis plays a key role. Obviously, this analysis needs to be started in the early phases of the project to catch the most benefits.

By returning to ground-related risks in construction and its industry, we would make a major step forward by reducing the life cycle costs of construction pro-jects. While this type of awareness appears to get ever more wide attention, a main obstacle to apply it in practice remains. It is the simple fact that, like design and construction, construction and maintenance are occasionally strictly separated responsibilities, at least with many public owned clients and government agencies. Several departments, with most of the time entirely different interests and object-ives, are not directly driven by a life cycle approach, because they do not gain any financial benefits from it. The increase in the application of modern design, construct and maintenance contracts, which include maintenance activities over 10 years, 20 years or even 30 years, is expected to give an enormous boost to life cycle thinking. Within such contractual arrangements the contractor can improve his long-term profits by pursuing the lowest total cost. With the right contractual incentives, a maximum fit-for-purpose can be achieved as well.

The GeoQ ground risks management approach may largely support further implementation of the life cycle approach in the construction industry. By its application we are able to concentrate on ground-related risk from the very beginning of the project, when we start in the feasibility phase. In the early stages we will identify the ground-related risk adversely affecting the life cycle costs and fit-for-purpose of the project. Typically, settlement of infrastructure is such a risk type. Let us imagine the effects on the operation and maintenance costs of a high-speed railway if, shortly after completion, unacceptable differential settlements surface, which do not allow a safe operation. It would be a nightmare for the involved parties.

Most likely many of us were not yet able to apply an entire life cycle management approach in recently completed projects. Still, we can create an opportunity to gain some of the financial benefits of the life cycle approach by looking forward through the project's entire operational life time and expected maintenance. It is possible to repeat the GeoQ steps of *gathering* project *information*, *risk identification* and *risk classification*, but now from a cost-effective operations and maintenance perspective. Next we can analyse the role of the ground conditions within these risks, for instance by a risk cause and effect analysis. I recall the tools as presented in Chapter 7, such as the fault tree analysis (FTA) and failure mode and effect analysis (FMEA). Based on the results, we would be able to schedule a ground-driven maintenance programme for the next 10 or 20 years. Preferably, we would consider a few scenarios, for instance with regard to the anticipated traffic intensity. An ongoing monitoring programme greatly assists the appraisal of long-term expected (ground) behaviour of the construction. Any deviations from the expected behaviour can be interpreted and may result in a cost-effective modification of the maintenance programme.

There is a lot of knowledge available about optimizing operations and maintenance programmes in the literature and in other sectors, such as the oil and gas industry. The life cycle management approach has, for instance, been largely developed and exploited in the energy sector.

In summary, the GeoQ process seems to be well-applicable in the maintenance phase of operation of realized construction projects. This may contribute to any party who is responsible for these activities or just using the project result. It facilitates cost-effective operation and maintenance by the structured and risk-driven consideration of the relevant ground data. These data need to be made available in a structured format, by risk registers and supporting databases. In the many projects where GeoQ could not yet have been applied before, it is still possible to start it in the maintenance phase.

Box 13.1 presents an example of a road renovation project, where GeoQ has been applied in a risk-driven contracting process, in search for a suitable maintenance contractor.

Box 13.1   Starting ground risk management in a road renovation project

An existing and busy road with daily traffic jams in the north-western part of The Netherlands urgently needed maintenance by renovation. The government client of the road decided to bring this renovation project in design and construct contract to the market, in order to gain the cost, time and quality benefits of a design that is well-integrated in the reconstruction works. A clear allocation of ground-related risk has been considered as necessary to meet these objectives. During construction of the road, many years ago, the GeoQ ground risk management process has not been available. Therefore, the client decided to start with the GeoQ process, right at the road's maintenance phase.

The first GeoQ step involved *gathering* project *information* on the three main objects of the project: embankments, viaducts and road-crossings by a sunken construction. Existing site investigation data have been interpreted, including an existing but rather generic risk register. Based on the available information, ground-related risks were identified in GeoQ step two and subdivided in geotechnical risks, geohydrological risks, geoenvironmental risks, and risks because of man-made obstructions in the ground. For the *classification* of all identified risks, GeoQ step three, the semi-qualititative method facilitated by the Electronic Board Room (EBR) has been used. Representatives of the client, the engineer and geotechnical consultants participated in this team-based risk session.

In total 19 geotechnical risks, 10 geohydrological risks, 5 geoenvironmental risks and 5 man-made obstruction risks were identified, classified and allocated to the client, the contractor or shared. The risk of deforming adjacent structures appeared to be one of the most serious ground-related risks of this typical greyfield project. As part of GeoQ step 4, four remediation measures were proposed for each ground-related risk. The preparation of a GBR, with clear baselines, was one of the recommendations to the client, while *evaluating* the risk profile by GeoQ step five. The results of the GeoQ process were stored in a risk register, concisely reported and presented to the client, which involved GeoQ step six of this project (GeoDelft, 2004).

The brief case study in Box 13.1 demonstrates it should never be too late to start a structured ground-related risk management process.

Rational risk management for existing dikes

In many areas of the world, cost-effective maintenance of *dikes* is of major safety concern for the people living behind them. This is particularly the case when the fit-for-purpose of a dike alters. Box 13.2 presents an example.

Box 13.2   Dike failure as a result of drought

To everyone's surprise in the very dry summer of 2003 in The Netherlands, two secondary dikes failed. Thousands of cubic metres of water flooded into domestic areas, creating property damage. The scale and effects of these two events were rather small, however, the conditions of these events raised large concern in society, as well as in the geotechnical community.

Thorough joint research by a number of Dutch parties resulted in a remarkable conclusion. It was not an extreme *wet* period that triggered the failure of these dikes, as is normally anticipated in dike design and construction, but an extreme *dry* period. This has been dramatically changing the strength properties of the dike material. These two dikes were typically not fit-for-purpose for such climatic conditions during their life time. This raised major public concern about the safety of the some remaining 17 000 km of secondary dikes that surround the many polder areas in the low lying parts of The Netherlands. What would happen if the dry period were succeeded by a serious wet period, with a lot of rain and consequently high water levels?

Because there will always be limits to maintenance budgets, a risk-driven approach may help to establish cost-effective dike maintenance programmes. Ideally, dike maintenance and reinforcement are restricted to those parts of the dike demanding it. As demonstrated by the example in Box 13.2, the conventional process approach of visual dike inspection, appraising the main failure mechanisms, providing strength and stability calculations and executing the resulting maintenance and reinforcement programmes may need support. Therefore, a new approach, to consider the failure risk of existing dikes throughout their entire life time, has been developed. This method, *Rational Risk management for Dikes* (RRD) calculates the failure risk of a particular dike section for a number of scenarios (Beetstra and Stoutjesdijk, 2005).

In The Netherlands and in many parts of the world, dikes are generally characterized by rather limited available ground data, which is logical because of the considerable lengths of these structures. The RRD method uses several *scenarios* to deal with this inherent incompleteness of ground information. These scenarios are established by combining a few geotechnical ground models, including ground layers and strength properties, with a number of geohydrological conditions, such as groundwater levels and external high-water conditions. In addition, cultural-historical information about the region and its inhabitants may help to assess the dike material properties. For instance, in The Netherlands, large areas have been exploited by the excavation of peat over the ages to use that material for heating.

As a result, dikes in those areas typically contain a lot of peaty material. The geometrical profile of the dike is also included in the scenarios. Figure 13.2 presents the development of eight scenarios for a dike.

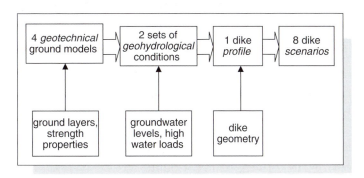

Figure 13.2: Scenarios for rational risk management of an existing dike.

The RRD approach determines a factor of safety against dike failure for each scenario by using conventional dike stability calculations. By appraising the regional *geological heterogeneity* and using geostatistics, the probability of occurrence of each scenario is calculated as well. This exercise results in the probability of occurrence of each scenario with its safety factor against dike failure for the dike sections of concern (Beetstra and Stoutjesdijk, 2005).

If the scenario with the highest probability of occurrence provides an unacceptably low safety factor against failure, according to the prevailing standards and codes for dike design, then there is reason for additional action. There is even more reason for concern if the majority of scenarios provide low safety factors. These situations clearly demand specific attention to that particular dike section, probably by performing a detailed and advanced ground investigation, to verify the ground properties as assessed in the scenarios. Based on this new information, decisions about the urgency for dike maintenance and reinforcement can be made on a largely *rational* basis, to assure an acceptable degree of dike safety during the remaining part of its life time. Obviously, it is recommended to repeat this procedure on a regular basis, for instance once every 5 or 10 years, to allow for any changes in external circumstances which may affect the pre-set dike safety. The expected main advantages of the application of the RRD method are:

- *Speed*: in a short period of time, within weeks or even a number of days, there is information available about the high risk parts of the dike; with a conventional approach, including field and laboratory investigations, this would have taken months at least

- *Cost-effectiveness*: identifying and improving the sections with the highest risk will result in a more cost-effective reduction of the overall risk of dike failure

- *Focus of risk-driven ground investigations*: detailed and advanced field and laboratory investigations are directed to the most vulnerable sections of the dikes, as a result of the RRD exercise

- *Presence of risk maps*: the RRD approach delivers risk maps that can be applied as a basis for crisis management in situation with extreme high water conditions. Because the weak spots are known, any risk effect measures, such as dike repair after (partial) failure and evacuation schemes, can be planned accordingly.

Obviously, this method can also be used to check the stability of many other structures, regarding probably most failure mechanisms of concern. We should, however, realize that this rather objective method of dike failure risk assessment includes some inherent subjectivity at the input side. For instance, the assessment of the probabilities of geological heterogeneity within the scenarios requires geological judgement and interpretation that stretches beyond the application of solely factual data. Also this rather quantitative risk assessment of dike safety tends to be a combination of hard or factual and soft or interpretative data. Nevertheless, the RRD approach is typically a member of the family of GeoQ-tools. It serves the maintenance of (very) long structures over large distance during their entire life time. Any foreseen changes in external conditions, which may challenge the fit-for-purpose of these structures over the years, can be incorporated into the scenarios. The first case study, later in this chapter, explains the application of the RRD method in more detail.

Ground dispute resolution after project completion

Despite the degree of ground risk management in all the preceding project stages, as well as during the maintenance phase itself, some inherent uncertainty of ground conditions remains. Therefore, also in the maintenance phase, the problem of differing site or ground conditions may rise (again). Figure 13.3 presents three main causes that may challenge the fit-for-purpose of a realized construction project during its operational life time.

A typical example of *different ground behaviour* is the often mentioned problem of unacceptable (differential) settlements. Even if settlement criteria are met at the moment of completing construction, unacceptable settlements may arise in the years to come. A much slower consolidation process caused by lower ground permeability than expected may for instance cause this problem. Also creep effects may be higher than anticipated.

If the traffic intensity on a road is significantly higher than expected during its life time, a *change of loading conditions* occurs. Or, in the case of an airport, if a new and larger type of aeroplane is operated. Loading conditions of offices or houses may increase, as part of a renovation programme. The latter is typically the case for historic buildings in a number of cities in The Netherlands, such as Amsterdam. These buildings, often several hundreds of years old, are typically founded on wooden piles. These gain a very low safety factor because the effect of negative skin friction, caused by ground settlement, was not been incorporated into pile design in those early days. As a result, even a minor change in the loading conditions may have an adverse effect on the critically loaded pile foundation. Serious damage is not unknown in these circumstances.

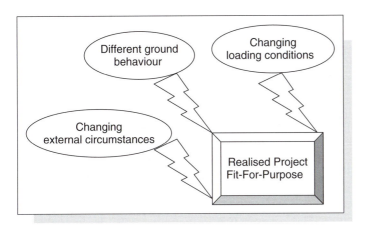

Figure 13.3: Challenges to the fit-for-purpose of realized projects.

With regard to the third change that may reveal different site conditions, a *change of external circumstances*, I build forward on the example of the wooden pile foundations. In many cities, groundwater levels fluctuate over time, for all kind of reasons like industrial activities, construction, excessive rain, or the opposite, excessive drought. However, if the top of a wooden pile becomes free of groundwater, due to a lowering of the groundwater level, fast deterioration of the pile occurs, with a serious risk for settlement damage or worse.

Obviously, also the combined effects of these three causes may become visible, which will significantly increase the problem's complexity. When encountering these types of problems during the life time of any kind of object, it will help to find appropriate solutions by distinguishing these causes of the adverse construction behaviour. It is always easy, but not always right, to point to the inherently uncertain ground, as the ultimate main source of the problem. Obviously, the ground could be the main cause, but not necessarily.

The main benefit of any available ground-related risk register, preferably with some supporting databases with ground information, is allowing rather objective verification of the role of any differing ground conditions. By comparing the actual ground conditions or behaviour with the available ground information and risk history, the existence of any differing ground conditions is likely to become clear. The continuation of at least some monitoring of ground behaviour after completion of construction supports any judgement about the actual ground conditions. If a GBR is available, it is, furthermore, much easier to check whether the differing site conditions are the responsibility of the client owner or (still) of the contractor. This approach will facilitate relatively fast and cost-effective, ground dispute resolution, according to the procedures as set out in Chapter 12. Obviously, the application of all these GeoQ ground risk management tools provides no guarantee that differing ground conditions can be totally avoided, nor their disputes.

Risk filing and mobilization by modern ICT tools

The sixth and final step of the GeoQ process has been defined earlier as the storage of all relevant ground risk information in risk registers, to be followed by its mobilization or transfer to the next project phase. Obviously, it is wise to keep a copy of the entire risk register, in order to benefit from it, perhaps even after several years, when some kind of liability may arise because of an adverse behaviour of the realized construction project .

There are numerous ways of presenting risk, on paper or digitally. Again, I advocate the KISS principle here; try to *Keep It Simple and Short*. The successors of our risk registers will be grateful to us when they receive any ground risk information in a decent and easily accessible way.

Modern ICT tools highly facilitate a structured storage of ground risk information. The results of a 1998-survey, as presented by Clayton (2001), about the availability of risk management software, presents more than thirty software packages, including their proposed applications and the addresses of their websites. Figure 13.4 presents these software packages covering, largely, four main categories.

Data management systems will be most suitable for risk filing in registers and its transfer. No doubt today a lot of updates and new software are available, compared with the 1998 search. A quick Internet survey will typically reveal a lot of up-to-date risk management software providers with their services. Any further attempt to present today's fast software developments in a book is obsolete before the book is printed.

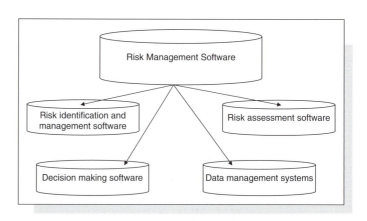

Figure 13.4: Four categories of risk management software.

Any available *Enterprise Resources Planning* (ERP) tools or software that is being used for *Knowledge Management* purposes may be used for risk management purposes as well, possibly with some modifications. In addition, experience teaches that the use of simple spreadsheet programmes largely support structured storage and use of ground risk management information, in particular for the smaller and less complicated types of construction projects. It will depend entirely on the type and size of the project, the software already available, as well as the taste of those involved in the ground risk management process, as to which software will be most cost-effective and efficient in its use.

Finally, once in the last GeoQ phase of maintenance in the operational life of the project, we might be able to benefit from these accumulated and software-supported risk registers, for instance, in the discussed case of perceived differing ground conditions, as ground uncertainty remains.

Case studies

This section presents two case studies in a rather brief format, because we approach the end of this book. These cases concern ground risk remediation during some sort of *maintenance* during the operational lifetime of realized projects. The purpose of these cases is to demonstrate how the GeoQ process may work in practice during the life time of construction projects. The first case describes the application of the RRD method for the safety assessment of an existing dike. The second case demonstrates the way GeoQ supports operational risk management at a waste disposal site. This latter case has a typical geoenvironmental focus and deals with the control of polluted groundwater.

Rational risk management approach for dike safety assessments

As part of a pilot project, the RRD method has been applied for some 80 km of dike in the low-lying and densely populated province of North-Holland in The Netherlands. The main objective was to obtain indicative risk-based information about the stability of the dike under wet loading conditions. A too low safety factor might result in dike failure and consequently a flood when a high water condition rules.

According to the first GeoQ step of *gathering information*, all existing ground-related data of the dike under consideration were collected. Next, as part of the *risk identification* step, the safety risk of the existing dike is considered in view of the prevailing codes and standards. This pilot project focused on the risk of slope instability of the dike under high-water loading conditions. Different from all other *risk classification* methods discussed in this book so far, next the risk of dike instability is assessed by a mainly *quantitative* approach. The dike has been subdivided into sections of 100 m length, which resulted in 800 sections, given the 80 km of dike. Based on the anticipated geological heterogeneity and the required degree of detail of the assessment, the length of these sections can typically be adapted for other projects.

For each dike section a number of scenarios has been defined, based on the expected geotechnical and geohydrological conditions, as discussed earlier. Any existing site investigation data proved to be of great help to assess the geotechnical and geohydrological properties for each scenario. Next, for each of the 800 dike sections and the related scenarios, the safety factors against slope instability were calculated, which was a highly automated process. Normally, some 3000 dike slope stability calculations can be easily performed in one night.

By geostatistical methods, which account for the expected geological heterogeneity, the calculated safety factors have been related to a probability of dike failure in each section. The resulting safety factors and their probabilities were grouped into three classes and plotted on a map. This was simply done by using red, yellow and green colours, indicating areas with an unsafe, moderately safe and safe situation against slope instability of the dike. This data set can be considered as the GeoQ *risk classification* of the slope instability of existing dikes. Figure 13.5 presents an example of a dike safety risk map. The dots should be considered as coloured red, yellow and green.

Figure 13.5: Safety factors of 800 dike cross sections (© with permission of GeoDelft).

Based on these results, largely rationally-based recommendations can be made for the next GeoQ step: the *remediation* of the slope instability risk of the most unfavourable parts of the existing dike. Strengthening of these unsafe parts by an extra embankment can be one of the risk cause remediation measures. Another option is to refine the risk assessment by additional ground information of the critical areas. A focused ground investigation can be performed in those areas that have been assessed as critical. As a result, boundaries of certain geological formations become known in more detail, ground properties can be up-dated and applied in revised scenarios. This may serve as input into a new cycle of stability calculations.

As part of the next GeoQ step of *risk evaluation*, the updated safety factor, as a result of the applied risk remediation measures, needs evaluation in view of the risk tolerance of the dike's stakeholders. Obviously, the risk perception of the people living around the dike is a factor of concern.

Because all data are digitally available, they can be easily *mobilized* as part of the last GeoQ step, towards any other stakeholders, such as a contractor, to assist the dike reinforcement activities or the government authority, who bears responsibility for the dike safety.

This case confirmed the foreseen advantages of the RRD-method, such as speed, cost-effectiveness and insight in the location of critical parts of the dike on accessible and easy to adapt risk maps (Pereboom et al., 2005).

While considering the inherent subjective element of the people factor, the RRD-approach is likely to become a welcome additional GeoQ tool. The applied software is available on the market and may also be achieved by linking existing databases and geotechnical calculation models. Finally, innovative readers may identify several other useful applications for this method, well beyond the dikes in this case.

Operational risk management at a waste disposal site

All together we create a lot of waste in our modern societies. In many areas of the world, waste is not only burned but also deposited. There is an increasing awareness that adverse chemicals may seep from the waste deposits into the underlying ground layers. Pollution of precious drinking water reserves may become a serious risk in domestic areas surrounding these waste deposit sites.

During the late 1990s, the concept of Flexible Emission Control (FEC) has been developed as a risk driven and dynamic approach to dealing with ground and groundwater pollution. Based upon the understanding of the ground and groundwater conditions, together with the characteristics of the pollutant, the pollution is controlled by the application of a monitoring system. Only if required, remedial actions are taken and the FEC-concept proved to be a cost-effective and efficient approach for managing risk caused by pollution (van Meurs et al., 2001). This implies the application of the observational method, in combination with monitoring, during the operational life time of any project or industrial activity that may cause some kind of environmental pollution.

The following case describes a risk-driven intervention during the operation of a waste disposal site in the northern part of The Netherlands. In the 1990s, a groundwater management system was installed on the site. In line with the FEC-concept, the main objective of this groundwater management system was to serve as a risk remediation measure. At that time, the identified risk of groundwater pollution, as a result of the leakage of contaminated materials from the waste deposit, was classified as serious. The site is underlain by permeable sand layers to a depth of about 80 m and drinking water is pumped from water-bearing layers in the neighbourhood of the site. Based on a sound groundwater management system, the waste disposal site was certified and was able to operate according to the Dutch government regulations.

The proper functioning of the groundwater management system is checked regularly by taking water samples from observation wells that have been installed around the waste disposal site.

At a certain moment, an increased chloride level was found in one observation well during regular monitoring. This situation might indicate the presence of so-called density currents in the subsoil, a phenomenon where the chloride contamination's own weight causes it to sink within the ground towards much deeper layers. Due to a groundwater flow in these deep layers, the chloride contamination could move towards the drinking water aquifer.

According to the GeoQ steps one to three, based on the monitoring and other relevant (ground) data, this risk was *identified* and *classified* as serious. In cooperation with the authorities, the manager of the waste disposal site decided to initiate an additional ground investigation, in order to map the presence and direction of movement of any chloride below and around the site. These data would be used as a basis to modify, where relevant, the groundwater management system, as a *risk remediation* measure by GeoQ step four. There was a high sense of urgency to this ground investigation, because adequate functioning of the groundwater system was an integral part of the licence to operate the waste deposit site. In practical terms, if the groundwater management system proved to be not working effectively, the authorities would be obligated to restrict dramatically or even to shut down the operation of the waste disposal site. Therefore, from a commercial point of view of the site manager, it was of utmost importance to adapt the system as soon as possible. As presented in Box 13.3, this situation demanded an innovative approach for a detailed ground investigation with advanced tools.

Because of the urgency of remediating the chloride contamination risk, the need for the advanced and dynamic ground investigation was of paramount importance and therefore widely accepted by all stakeholders. The applied procedures agreed well with the recommendations of the Network for Contaminated Land in Europe (NICOLE) (Network for Contaminated Land in Europe, 2002). The results of the ground investigation were applied in a geoenvironmental model study, which resulted in the required modification of the groundwater management system being the *risk effect remediation measure*. Ongoing monitoring will serve as the GeoQ step of *risk evaluation*, while the data will be reported on a regular basis to the parties involved, which acts as the final GeoQ step of *risk mobilization*.

The applied risk-driven GeoQ approach during the operational phase of the waste disposal site resulted in three months of time saving, which was very welcome in view of the site's commercial operations. This case serves as an example of how the GeoQ ground risk management approach may facilitate *geoenvironmental* type of problems during operation and maintenance of industrial activities.

Box 13.3   The benefits of an innovative environmental ground investigation

In order to gain the necessary additional ground related data as quickly as possible, innovative ground investigation tools were applied. Advanced equipment for deep cone penetration tests (CPT), including in-situ conductivity measurements were successfully applied to map the chloride contamination under and around a waste disposal site. Groundwater samples were taking by using the so-called *multi-groundwater sampling probe*, to depths of about 80 m.

This approach took considerably less time than the usual method for mapping pollution, which requires motoring of observation wells that need first to be installed around the contaminated location. Boring and installing one of these deep observation wells require at least five working days. Once in position, it takes another week of waiting for the subsoil conditions to return to their more or less initial in-situ conditions before a groundwater sample can be retrieved and analysed. Therefore, at least some weeks pass before there is any insight into the presence and flow direction of the contaminant.

The deep cone penetration test with conductivity measurements and groundwater sampling at several depths were performed in less than a day per location. It also provided immediate qualitative information about chloride levels. This made it possible to select the next site investigation location in order to map the chloride contamination efficiently. To guarantee the measured conductivity corresponding with the interpreted chloride levels, calibration of the groundwater samples with their conductivity was performed at several locations and depths. The chosen flexible and dynamic site investigation approach proved to be faster, cheaper and better than a more conventional approach by boring, installing monitoring wells, taking samples and providing laboratory analysis (van Meurs et al., 2001). The presented approach transformed this conventional sequence of four steps to mainly one step on site, supported by a number of laboratory calibrations.

Summary

This chapter has demonstrated the application of the GeoQ process in the maintenance phase of construction projects and even industrial operations. Smooth operation is the proof of effective ground risk remediation during design and construction. In this final stage, which may take 10 years, 20 years, 50 years or perhaps even 100 years, the ultimate GeoQ objective is to realize the project's most cost-effective maintenance, from a ground risk management perspective. Preferably, all of the six GeoQ steps are performed in order to identify, classify and manage any changes in the perceived and ground-related residual project risks.

Four ground risk management approaches, which are of particular importance in the maintenance phase, have been introduced and discussed. First, the *life cycle concept* is extended from a mainly cost perspective to a lifelong fit-for-purpose perspective. Any project should remain fit-for-purpose, even in case of unexpected and adverse ground behaviour, which can be triggered by changes in loading conditions or other external conditions over time.

Second, flood control is of major safety concern in many areas in the world and the fit-for-purpose of existing dikes need regular checks. This may reveal the need for maintenance or even reinforcement. The method of *Rational Risk management for Dikes* (RRD) calculates the safety factor against dike failure by using several scenarios, which proved to be a highly effective and efficient GeoQ tool.

Third, some (minimal) inherent uncertainty of ground conditions remains even during operation and maintenance. *Ground dispute resolution*, even after project completion, is not unthinkable. Differing ground conditions may have an adverse effect on the project's anticipated maintenance programmes. Continuous monitoring, even after project completion, may remain important for projects with high risk profiles, to obtain information on (different) ground behaviour, which can be used in case of disputes about differing ground conditions. These three ground risk management tools are related to the fifth GeoQ step of *risk evaluation*.

The last and fourth ground risk management approach for the maintenance phase concerned a quick appraisal of modern information and communications technology (ICT). Of four main risk management supporting software types introduced, data management systems will probably be the most suitable for filing risk in risk registers and its transfer, which concerns the sixth GeoQ step of *risk mobilization*.

Two case studies with risks of geotechnical, gehydrological and geoenvironmental origin illustrated the possibilities of GeoQ ground-related risk management in the practice of operation and maintenance.

PART FOUR

A look into the future

14 *To end with a new start*

Introduction

This final chapter highlights briefly some of the main opinions and conclusions of this book. It starts with the soft systems or people aspects of ground risk management and is followed by the process of applying the GeoQ ground risk management framework, which combines the soft systems with the hard systems of technology. The last section of this chapter presents a brief outlook forward towards a prosperous construction industry, as perceived from a ground risk management perspective. I end with some last words.

Ground risk management: the people

First of all and probably most importantly, due attention to the people factor is considered to be the key success factor for a cost-effective and efficient operation of ground risk management. We need both individuals and teams. Representatives of all stakeholders in any construction project should preferably reach a shared basis of understanding of hazards, risk and risk management concepts and practices. This understanding should stretch from the traditional reactive risk approach towards the more modern proactive and holistic approaches, which include the social construction of risk and its associated difficulties. Risk management should not solely be perceived as an unavoidable vehicle to avoid crises, but as an overall way of proactive management of risk and opportunities. Management of ground-related risk calls for acknowledgement by all construction players for its short-term and long-term bottom-line benefits, which concern not only financial, but also social and environmental aspects.

At some time, any risk-driven change agent will probably need a substantial portion of individual motivation and dedication to convince the project stakeholders about the necessity and benefits of structured ground risk management. Therefore, we have explored the concept of the individual and our individual contribution to construction projects. Effective ground risk management starts with an awareness of the inherent differences in individual risk perception.

This focus on the individual is a typical aspect of the dominating cultures in the western world. However, no individual can create any construction just solo. We need (many) other people and we need both mono-disciplinary expert teams and multidisciplinary teams. These project teams need a proactive risk culture, with continuing attention to people aspects, such as organizational processes, behaviour, culture and real participation of all involved employees. Also our interaction with the people beyond the team, clients and the public in our society, has been thoroughly discussed. We encountered elements of collectivism, which is a characteristic of many cultures in the eastern part of the world.

We need both approaches, individualism and collectivism, for cost-effective and efficient risk management in today's increasingly globalizing world. The combined individual and collective willingness of at least the key project players is invaluable, because these people make it happen. Our challenge is to find the optimum balance between both approaches.

Ground risk management: the process

Ground risk management is entering the construction arena in order to control, reduce or even eliminate foreseeable ground-related risk. Moreover, the dark and unknown ground is no longer only a factor of risk, ground-related opportunities are caught as well. The GeoQ process is nothing more than a simple and adaptive framework for facilitating this process. GeoQ is based on several existing and widely accepted risk management approaches that can be found in many text-books, all over the world. The cyclic character, by six generic risk management steps that should ideally be performed in each project phase, helps to rationalize, to structure, to minimize *and* to communicate potential major ground-related problems. Despite the simplicity of the concept, GeoQ is rather difficult to implement because of the inherent presence of the people factor in risk management.

To maximize the financial benefits, GeoQ should *start* as early as possible in any construction project and be *continued* as long as possible, even including the maintenance phase during operation, *after* completion of construction. GeoQ provides a direct relationship between major technical and ground-related project risk and the required type, quantity and quality of ground expertise and investigations for the risk's remediation. Therefore, the valued ground expertise and investigations become explicitly linked to the remediation of project risk.

The cost of good-quality ground-related services during engineering and construction is usually only a fraction of the cost of a major risk event, as well as of the cost saving by optimizations and value engineering. In his John Mitchell Lecture paper, Heinz Brandl (2004), professor at the University of Vienna, Austria, presents the following quote of the British social reformer John Ruskin (1819–1900):

> There exists hardly anything in this world that could not be produced in a lower quality and be sold at a lower price – and people who orientate themselves on the price only are the natural prey for such practices. It is not clever to pay too much, but it is even less clever to pay too little. When you pay too much, you lose some money – that's all. When, on the other hand, you pay too little, you sometimes lose everything, because the purchased object cannot fulfil its intended purpose. The law of economics does not allow to obtain big value for little money. If you take the lowest offer, you have to add something for the risk you take. And if you do that, you also have enough money to pay for something better than the lowest offer.

The application of the GeoQ process helps to communicate and justify the need for adequate ground investigations and related geotechnical, geohydrological and geoenvironmental services to the sometimes less-technically-oriented decision-makers in construction projects and society.

In addition, besides risk management, GeoQ supports several other management approaches, such as quality management, value management and innovation management. Partly as a result of this hybrid character, GeoQ applications in a variety of large and small projects demonstrate considerable reductions of cost and time, as well as minimised individual and team frustration. The GeoQ process proves to reduce ground-related risk to accepted levels and to exploit ground-related opportunities as well.

To gain (even more) financial and other benefits, I recommend at least considering the following aspects for any construction project:

- Applying a well-structured ground risk management approach, for instance by the proposed and proven GeoQ process

- Establishing a proactive and risk aware culture within the project teams, by combining the best of the individual and the team approach, with due attention to the people factor

- Providing an awareness, at least qualitative and preferably more quantitative, of the cost of major risk events, as well as the cost of risk remediation measures and their management

- Selective use of risk management tools: adequate techniques should be selected in each project phase in order to control, reduce or even eliminate the most critical ground-related project risks.

Based on both individual and team-based risk awareness, as well as the willingness to deal with risk beyond its avoidance, these recommendations provide some keys to make a difference in our practices.

Towards a prosperous construction industry

In today's construction industry we have to deal with a lot of change. The following statement by Todd D. Jick (1993), who taught at Harvard Business School, may be recognized by many readers: 'Changing is inherently messy, confusing and loaded with unpredictability. No one escapes it'. When investigating ground, we are used to getting our hands dirty. Ground is inherently unpredictable and, while constructing, we cannot escape the ground, whether we like it or not. Change and ground have apparently a lot in common.

It will take some more years at least, perhaps even a generation, until the presented GeoQ ground risk management framework, or similar approaches, are fully accepted and embedded in our day-to-day engineering and construction activities. However, I think we are in some kind of a hurry, because many problems in the construction industry, and society as a whole, need fast relief. Given this situation, how could and should we operate in our day-to-day practices?

Therefore, particularly related to the time factor, transformation within the construction industry deserves some more attention. In my opinion, industry transformation is a rather continuous process of adapting to changes in society, or even better, anticipating the expected demands of society. Herman Wijffels (2004), former chairman of the Dutch Social-Economic Advisory Board of the Dutch Government states: 'At this moment we are in a transition process from a mechanistic order towards an organic order'. Many consultants apply so-called *evolutionary* processes in attempting to realize organizational and industry transformation. By the evolutionary process it is expected that a transformation will take a rather long period of time, 5 years, 10 years, perhaps even longer, before the transformation is really completed, provided that any transformation will end at any stage. This provides a step-by-step approach, contrary to the more *revolutionary* or shock and awe way of change.

In this respect, I was rather surprised to read about *organic* evolution, as occurring in nature, which is apparently not always a smooth and step-by-step like process. Even periods of radical change belong to an evolutionary process (Marx Hubbard, 2004). Box 14.1 presents a few characteristics of organic change that I have selected, because these may help us better to understand the transformation processes in our organizations and industry. I derived the aspects from Marx Hubbard (2004) and interpreted them from a risk-driven construction perspective. I added the sixth aspect myself.

Without usurping evolutionary biologists, the issues presented Box 14.1 seem to confirm that construction industry transformation is a natural process. This is similar to many other transformation processes around us, including our own human evolutionary development process. An evolutionary process is not necessarily slow and dull, as the lessons of nature teach us, which may stimulate us

Box 14.1 Similarities of organic and industry transformation

The following aspects of organic change may help us better to understand the transformation processes in our organizations and industry:

1 Evolutionary transformations are in fact *quantum transformations*, a jump from one condition to another, instead of a smooth step-by-step growing process to another condition. In fact, in our modern times the Internet created such a quantum leap in society. I foresee a similar jump in our construction industry, when modern ICT and Internet technologies become fully exploited. If, for instance, the ICT-facilitated observational method, supported by on-line monitoring becomes common practice, our entire design and construction processes will radically change.

2 Real transformation is always the result of a *crisis*. When nature reaches its boundaries, there is not automatically stabilization at that boundary, but a process of innovation and transformation starts. Following the previous discussion about the challenges and opportunities in the construction industry, in my opinion some sort of crisis is ongoing in the global construction industry.

3 *Holism* is part of nature and *integration* is part of the evolutionary progress, which aligns with the suggested integration of the hard technological and soft human systems, in construction in general and particularly in ground-related risk management.

4 Evolution creates *beauty* and only beauty survives. In nature, beauty is often related to strength and fertility. Its objective is to guarantee survival. Although beauty is a very subjective topic, construction has created a lot of beauty since ancient times. Many impressive man-made structures have survived hundreds or even thousands of years. I like the idea of bringing the concept of beauty, as a vehicle for quality over time, to a more prominent position in the construction industry.

5 Evolution increases *awareness* and *freedom*, which is clearly expressed by our own evolution in the (post-) modern era. Our awareness about the surrounding world and our freedom of choice has dramatically increased over the last decades. Again, the Internet and modern ICT are huge catalysts. This increasing awareness and freedom of choice of billions of people, all over the world, will change their expectations of the construction industry. By working in that industry, it is our responsibility to deliver fit-for-new-purpose products and services, in order to satisfy the new needs of our clients and to make a decent profit as well.

6 Evolution is *here* and *now*, like risk, organizational change and industry change, no one can escape it.

to take bigger steps in our transformation processes. We should dare to make even leaps if required in certain situations, while still remaining evolutionists rather than revolutionists. This approach is quite different from the conventionally advocated and well-established prudent and often reserved step-by-step-approach of change. Let us try to act differently in order to embed the concept of structured ground risk management as effectively as possible in our (project) organizations and industry.

When it is difficult to push forward with any change initiative, I often remember the small dharma puppets, which I bought for my little daughters some years ago in Tokyo. These puppets do not have arms and legs, which is apparently not convenient. However, due to this handicap, they always come back to an upright position, however large and different the forces on them are. We should act as dharma puppets if we feel handicapped to bring the change we think that is needed in our projects.

In this respect the psychologist Martin E.P. Seligman inspires me. After many years of being a rather conventional psychologist, basically involved with the *curative* aspects of psychology, he developed a new psychological approach. This concerns mainly avoiding psychological problems, in a *preventive* way. Prevention has a much more certain outcome than trying to cure. He became the pioneer of the *Positive Psychology Movement* (Seligman, 2002).

Analogous to this development in psychology, I am convinced that it is our time to develop a positive movement in the construction industry, with a leading role for ground engineering. In my experience, ground engineering is still applied in too much of a curative way. It has a rather negative reputation as a source of problems with often fuzzy solutions, in which many experts have totally different opinions. Therefore, let us apply risk management to transform ground engineering and its related disciplines from a mainly curative towards a much more preventive approach. Let us start some kind of positive movement of ground engineering as well as a catalyst for the entire construction industry. In this book I have presented a lot of thoughts, opinions, tools and experiences to start to make a move in that direction. I wish you all kinds of success in making a positive difference!

Some last words

Finally, once you have read this book, I want to recall and share some of the thoughts and intentions I raised in the introduction of this book. My message with this book is to *always look at the bright side of risk*. There is a lot of negativism about the concept of risk, but any risk we imagine is some fruit of our own imagination. This helps us to get at least some minor insight into the many possible futures we might experience. By identifying a foreseeable risk, at least we realize something unfortunate that can happen. So long as it is a risk, it has not yet effectuated and we are still able to do something with it, by reducing its probability, reducing its adverse effects or a combination of both ways of action. Furthermore, often risks hide surprising *opportunities*. This is what I call the *bright side of risk*.

I aim to create a much more positive image around the concepts of risk and risk management in general and ground-related risk management in particular. With this book I have tried to communicate the benefits of encountering risk and acting upon it in an effective way, rather than the conventional, very human impulse of risk aversion. My approach is to capture foreseeable risk within our zone of influence, or in the zones of influence of our project team, our client, our industry and our society. The remaining risks, the ones outside our circles of influence and the unforeseeable risks, those we have to accept. It should be possible to accept them, if we are sure that we tackled all their foreseeable colleagues, individually or by joint forces. This book aims to give a contribution to this innovative approach and application of ground-related risk.

A last remark to stay alert. So long as construction is man-made and deals with *Mother Earth*, ground-related risk remains in our day-to-day reality. Ground-related problems and even crises will continue to happen. However, hopefully, an increasing acceptance and application of structured ground risk management in the world's construction industry can reduce the number, intensity and impact of these ground problems and crises. The future will teach us about the outcome of the suggested developments and their ultimate added value, preferably expressed in sound sums of money.

References

Abbott, E.L. (1998). Preparation of contract documents for subsurface projects. In *Subsurface Conditions: Risk Management For Design and Construction Management Professionals* (D.J. Hatem, ed.) pp. 95–128. John Wiley & Sons, New York.

Abramson, L.W., Cochran, J., Handewith, H. and MacBriar, T. (2002). Predicted and actual risks in construction of the Mercer Street Tunnel. In *Proceedings of the North American Tunneling Conference*, May, Seattle, pp. 211–18. Balkema, Lisse.

Altabba, B., Einstein, H. and Hugh, C. (2004). An economic approach to risk management for tunnels. In *Proceedings North American Tunneling 2004* (Ozdemir, ed.) pp. 295–301. Taylor & Francis Group, New York.

Ameratunga, J., Shaw, P., Beohm W.J. and Boyle, P.J. (2005). Seawall construction in Moreton Bay, Brisbane. In *Proceedings 16th International Conference on Soil Mechanics and Geotechnical Engineering*, 12–16 September, Osaka, Japan, pp. 1439–42. Millpress, Rotterdam.

American Society of Civil Engineers (1980). *Construction Risks and Liability Sharing, Volume II.* ASCE, Reston.

Anderson, D.R., Sweeny D.J. and Williams T.A. (1999). *Statistics for Business and Economics*, 7th edn. South-Western College Publishing, Cincinnati.

Ansoff, H.I. (1984). *Implanting Strategic Management.* Prentice Hall, Englewood Cliffs.

Association of Engineering Firms Practicing in Geosciences (1988). *Alternate Dispute Resolution for the Construction Industy.* ASFE, Silver Springs.

Aufenhanger, J. (1985). *Philosophy* (Dutch edn.). Het Spectrum, Utrecht.

Australian Geomechanics Society (2000). Landslide risk management concepts and guidelines. *Australian Geomechanics Journal*, **35**(1), 51–92.

Barends, F.B.J. (2005). Associating with advancing insight: Terzaghi Oration 2005. In *Proceedings 16th International Conference on Soil Mechanics and Geotechnical Engineering*, 12–16 September, Osaka, Japan, pp. 217–48. Millpress, Rotterdam.

Barends, F.B.J. and Mischgofsky, F.H. (2005). European innovation programmes for urban infrastructure. In *Proceedings International Symposium on Urban Geotechnics*. GeocityNet2005, Lille.

Baya, D., Oluwoye, J. and Lenard, D. (1997). An analysis of contractor's approach to risk identification in New South Wales, Australia. *Construction Management and Economics* **15**, 363–9.

Beetstra, G.W. and Stoutjesdijk, T.P. (2005). *First Approach to Rational Risk Management (RRD) by the Delta Institute*, 1 November. GeoDelft, Delft.

Bell, F.G. (1987). *Ground Engineer's Reference Handbook*. Butterworth and Co Publishers, London.

Bellotti, R. (1989). Shear strength of sands from CPT. In *Proceedings 12th International Conference on Soil Mechanics and Foundation Engineering*, August, Rio de Janeiro, pp. 179–84. Balkema, Rotterdam.

Berlo, K. (1960). *The Process of Communication: An Introduction to Theory and Practice*. Holt, Rinehart and Winston, New York.

Bijsterveld, K. (2005) How do you come to client focussed construction? (in Dutch). *Building Business*, **1**, February, 62–3.

Bles, T.J. (2003). *Risk During the Installation of Piled Foundations: Decision Support System by Identifying, Classifying and Balancing Risk* (in Dutch). MSc Thesis, Twente University of Technology, Enschede.

Bles, T.J., Hemmen, B.R. and van Staveren, M.Th. (2005). *Dealing with Risks using GeoQ and GeoBrain*. User Forum Risk Management in System Engineering Life-Cycle, 6 and 7 October, Gesellschaft für Systems Engineering, Bremen.

Block, P. (2002). *The Answer to How is Yes: Acting on What Matters*. Berret-Koehler Publishers, San Fransisco.

Blockley, D. and Godfrey, P. (2000). *Doing It Differently: Systems for Rethinking Construction*. Thomas Telford Ltd, London.

Bock, H., Blümling, P. and Konietzky, H. (2005). Common ground in engineering geology, soil mechanics and rock mechanics: past, present and future. In *Proceedings of the Symposium New Developments in Geo Information Handling for Engineering Geology*, 21 January 2005, on the occasion of the retirement of IAEG President Niek Rengers, Synopsis p. 2. ITC, Enschede.

Boothroyd, C. and Emmet, J. (1996). *Risk Management: A Practical Guide for Construction Professionals*. Witherby, London.

Brandl, H. (2004). *The Civil and Geotechnical Engineer in Society: Ethical and Philosophical Thoughts, Challenges and Recommendations*. The Deep Foundations Institute, Hawthorne.

Brierly, G.S. (1998). Subsurface investigations and geotechnical report preparation. In *Subsurface Conditions: Risk Management for Design and Construction Management Professionals* (D.J. Hatem, ed.) pp. 49–94. John Wiley & Sons, New York.

Brierly, G.S. and Hatem, D.J. (eds) (2002). *Design Build Subsurface Projects*. Zeni House Books, Phoenix.

British Tunnelling Society (2003). *The Joint Code of Practice for Risk Management of Tunnel Works in the UK*. BTS, London.

Burland, J. (1995). Assessment of risk and damage to buildings due to tunnelling and excavation. In *Proceedings First International Conference on Earthquake Geotechnical Engineering*, November, Tokyo, pp. 1189–201. Balkema, Rotterdam.

Calle, E.O.F. (2002). Optimisation of ground investigation requires a clear probability analysis (in Dutch). *Land + Water*, **11**, 36–9.

Cameron, K.S. and Quinn, R.E. (1998). *Diagnosing and Changing Organisational Culture.* Addison-Wesley, Reading, Massachusetts.

Capra, F. (1983). *The Turning Point: Science, Society and the Rising Culture.* Bantam Books, New York.

Carlsson, M., Hintze, S. and Stille, H. (2005). On risk management in large infrastructure projects. *Proceedings 16th International Conference on Soil Mechanics and Geotechnical Engineering,* 12–16 September, Osaka, Japan, pp. 2785–8. Millpress, Rotterdam.

Caspe, H.P. (1998). Dispute resolution mechanisms for differing site conditions claims. In *Subsurface Conditions: Risk Management for Design and Construction Management Profesionals* (D.J. Hatem, ed.) pp. 223–57. John Wiley & Sons, New York.

Clayton, C.R.I. (ed.) (2001). *Managing Geotechnical Risk: Improving Productivity in UK Building and Construction.* The Institute of Civil Engineers, London.

Construction Industry Institute (1994). *Pre-Project Planning: Beginning the Project the Right Way.* Publication 39-1. CII, Austin.

Construction Industry Research and Information Association (1978). *Tunnelling: Improved Contract Practices.* CIRIA, London.

Construction Task Force (1998). *Rethinking Construction.* DETR, London.

Courtney, H., Kirkland, J. and Vigueri, P. (1997). Strategy under uncertainty. *Harvard Business Review,* November–December, 67–79.

Covello, V.T., McCallum, D.B. and Pavlova, M.T. (eds) (1989). *Effective Risk Communication: The Role and Responsibility of Government and Non-Government Organizations.* Plenum Press, New York.

Covey, S.R. (1992). *The Seven Habits of Highly Effective People.* Simon and Schuster, New York.

Cummings, D. and Kenton, F.J. (2004). Eleven case studies of failures in geotechnical engineering, engineering geology, and geophysics: how they could have been avoided. In *Proceedings: Fifth International Conference on Case Histories in Geotechnical Engineering, 13–17 April, New York* (S. Prakash, ed.) Paper No. 7.01, pp. 1–12. University of Missouri-Rolla, Rolla.

CUR (2003). *Determination of Geotechnical Parameters: Publication 2003–7* (in Dutch). CUR Foundation, Gouda.

CUR (1997). *Probabilites in Civil Engineering,: Part 1 – Probabilistic Design in Theories* (in Dutch). CUR Foundation, Gouda.

Daft, R.L. (1998). *Organisation Theory and Behaviour,* 6th edn. South Western College Publishing, Cincinnati.

Dean, E.B. (1998). *Total Quality Management from the Perspective of Competitive Advantage,* URL: http://mijuno.larc.nasa.gov/dfc/tqm.html.

de Bono, E. (1998). *Simplicity.* Penguin Books, New York.

de Gitirana, G.F.N. and Fredlund, D.G. (2005). The application of unsaturated soil mechanics to the assessment of weather-related geo-hazards. In *Proceedings 16th International Conference on Soil Mechanics and Geotechnical Engineering,* 12–16 September, Osaka, Japan, pp. 2515–20. Millpress, Rotterdam.

de Ridder, H.A.J. (1998). Dealing with risks in foundation engineering (in Dutch) *Geotechniek,* October, 49–51.

Den Haan, E.J. (1994). *A Simple Compression Model for Non-Brittle Soft Clays and Peat.* PhD Thesis. Delft University Press, Delft.

Dewey, J. (1927). *Human Nature and Conduct: An Introduction to Social psychology.* H. Holt and Co., New York.

Dibb, S., Simkin, L., Pride, W.M. and Ferrel, O.C. (1997). *Marketing: Concepts and Strategies,* 3rd European edn. Houghton Mifflin Company, Boston.

Douglas, W.S. (1974). Role of specifications in foundation construction. In *ASCE Journal of the Construction Division,* **100**, 199–201.

Dunniclif, J. and Green, G.E. (1988). *Geotechnical Instrumentation for Monitoring Field Performance.* John Wiley & Sons, New York.

Dunniclif, J. and Powderham, A. (2001). Recommendations for procurement of geotechnical instruments and field instrumentation services. *Geotechnical Views,* **19**(3), 30–5.

Dutch Construction Steering Committee (2005). *Annual Report 2004: From Preparation To Execution* (in Dutch). De Vries, Zierikzee.

Dykstra, C.J. and Joling, A.G. (2001). Practical value of coefficient of consolidation determined by Asaoka Method (in Dutch). *Geotechniek,* **5**(2), 104–10.

Edwards, L. (1995). *Practical Risk Management in the Construction Industry.* Thomas Telford, London.

Edwards, P.J. and Bowen, P.A. (2005). *Risk Management in Project Organisations.* Butterworth-Heinemann, Oxford.

Elliot, D., Letza S.R., McGuiness, M. and Smallman, C. (2000). Governance, control and operational risk: the Turnbill effect. *Syllabus Crisis Management,* NIMBAS University, Utrecht.

ENV 1997-1 (1994) *Eurocode 7: Geotechnical design Part 1: General rules.* CEN, European Committee for Standardisation, Brussels.

Essex, R.J. (2003). Design-bid-build contracting: climbing out of a paradigm sinkhole. In *Proceedings Rapid Excavation and Tunneling Conference.* Society of Mining, Metallurgy and Exploration, New Orleans.

Essex R.J. (ed). (1997). *Geotechnical Baseline Reports for Underground Construction.* Technical Committee on Geotechnical Reports of the Underground Technology Research Council, ASCE, Danvers.

European Construction Technology Platform (2005). *Challenges and Developments for the Built Environment in Europe: The Harmonised Vision for 2030 for a Sustainable and Competitive Construction Sector.* ECTP, Brussels.

Finch, A.P. and Patterson, R.L. (2003). Recent trends in procurement of tunnel projects in the United Kingdom. In *Proceedings Rapid Excavation and Tunneling Conference.* Society of Mining, Metallurgy and Exploration, New Orleans.

Flanagan, R. and Norman, G. (1993). *Risk Management and Construction.* Blackwell Scientific Publications, Oxford.

Fookes P.G., Baynes, F.J. and Hutchinson, J.N. (2000). Total geological history: a model approach to the anticipation, observation and understanding of site conditions. In *Proceedings International Conference on Geotechnical and Geological Engineering, EngGeo2000,* 19–24 November, Melbourne, Australia, Vol. 1, pp. 370–460. Technomic, Basel.

GeoDelft (2004). *Reconstruction of Motorway N242 Around the City of Alkmaar: Risk Identification from a Ground Perspective* (in Dutch). GeoDelft, Delft.

GeoDelft (2000). *Strategic Plan 2001–2004.* GeoDelft, Delft.

Gilmartin, T.F. (1998). Insurance for subsurface projects. In *Subsurface Conditions: Risk Management for Design and Construction Management Professionals* (D.J. Hatem, ed.) pp. 417–44. John Wiley & Sons, New York.

Glass, P.R. and Powderham, A.J. (1994). Application of the observational method at the limehouse link. *Géotechnique*, **44**(4), 665–97.

Godfrey, P.S. (1996). *Control of Risk: A Guide to the Systematic Management of Risk from Construction*. CIRIA Special Publication 125. CIRIA, London.

Goffee, R. and Jones, G. (2005). Managing authenticity: the paradox of great leadership. *Harvard Business Review*, December, 86–94.

Goleman, D. (1996). Emotional Intelligence (Dutch edn). Pandora Pockets, Amsterdam.

Gould, J.P. (1995). Geotechnology in dispute resolution – the twenty-sixth Karl Terzaghi lecture. *Journal of Geotechnical Engineering*, **121**(7), 523–34.

Grant, R.M. (1998). *Contemporary Strategy Analysis*. Blackwell, Oxford.

Gratton, L. (2004). *The Democratic Enterprise*. Prentice Hall, London.

Greeuw, G. and Van, M.A. (2003). Local data sets provide realistic strength parameters (in Dutch). *Land + Water*, **6**, 38–9.

Hamel, G. and Prahalad, C.K. (1994). *Competing for the Future*. Harvard Business School Press, Boston.

Handy, C. (2002). *The Elephant and the Flea*. Harvard Business School Press, Boston.

Hannink, G., Vrouwenvelder, A.C.W.M., Lindenberg J. and Calle, E.O.F. (2004). From uncertainty towards reliability: between code and practice (in Dutch). *Geotechniek*, **5**, 14–21.

Hatem, D.J. (1998). Professional liability and risk allocation – management considerations for design and construction management professionals involved in subsurface projects. In *Subsurface Conditions: Risk Management for Design and Construction Management Professionals* (D.J. Hatem, ed.) pp. 223–57. John Wiley & Sons, New York.

Hatem, D.J. (1996). Geotechnical baselines and geotechnical reality; one and the same, similar or not very close – professional liability implications. *The CA/T Professional Liability Reporter*, **2**(2), 1–8.

Hawkins, D.R. (1995). *Power vs. Force*. Hay House, Sydney.

Hedges, A. (1985). Group interviewing. In *Applied Qualitative Research* (R. Walker, ed.). Gower Publishers, Aldershot.

Heijboer, J., Makkinga, L.C., Smid, L. and Bredenoord, J. (2004). The contract and project managment. In *The Westerschelde Tunnel: Approaching Limits* (J. Heijboer, J. van den Hoonaard and F.W.J. van de Linde, eds) pp. 261–78. A.A. Balkema, Lisse.

Heijbrock, F. (2005). Construction worldwide in transition (in Dutch). *Cobouw*, 23 March, 1.

Herbschleb J., van Staveren M.Th. and Teunissen E. (2001). The North-South Metroline: geotechnical risk allocation by the Geotechnical Baseline Report (in Dutch). *Geotechniek*, **2**, 83–9.

Hertz, N. (2001). *The Silent Takeover: Global Capitalism and the Death of Democracy*. William Heinemann, London.

Hicks, M.A. and Samy, K. (2002). Reliability-based characteristic values: a stochastic approach to Eurocode 7. *Ground Engineering*, December, 30–4.

Hillson, D. (2002). *Risk Management Maturity Level Development*. Risk Management Specific Interest Group, Project Management Institute. Newton Square, Pennsylvania.

Ho, K., Leroi, E. and Roberds, B. (2000). Quantitative Risk Assessments: application, myths and future direction. In *Proceedings International Conference on Geotechnical and Geological Engineering, EngGeo2000*, 19–24 November, Melbourne, Australia, Vol. 1, pp. 269–312. Technomic, Basel.

Hoek, E. and Palmieri, A. (1998). Geotechnical risks on large civil engineering projects. In *Proceedings 8th Congres IAEG*, Vol 1, pp. 79–88. A.A. Balkema, Rotterdam.

Hölscher, P. (2003). *Influence of Monitoring on Reliability of Settlements Predictions: the Application of the Isotache model* (in Dutch). Delft Cluster Report 01.01.07-07. Delft Cluster, Delft.

Hounjet, M.W.A. (2005). *GeoCheck: Urban Development for the Almere Pampus Project* (in Dutch). GeoDelft, Delft.

Imai, M. (1986). *Kaizen: The Key to Japan's Competitive Success*. Random House, New York.

Jansen, C.E.C. (2001). *Completion and Content of Design and Construct Contracts for Complex Infrastructural Projects* (in Dutch). Kluwer, Deventer.

Jick, T.D. (1993). *Managing Change: Cases and Concepts*. Irwin/McGraw-Hill, Boston.

Jobs, S. (2005). *You've Got to Find What You Love*. Stanford Report, June 14. Stanford University, Stanford.

Johnson, G. (1992). Managing strategic change: strategy, culture and action. *Long Range Planning*, **25**(1), 28–36.

Katzenbach, J.R. and Smith, D.K. (1994). *The Wisdom of Teams: Creating the High-Performance Organisation*. Harvard Business School Press, Boston.

Keizer, J.A., Halman, J.I.M. and Song, M. (2002). From experience: applying the Risk Diagnosing Methodology. *The Journal of Product Innovation Management*, **19**, 213–32.

Kelly, K. (1996). The new biology of business. In *Rethinking the Future* (R. Gibson, ed.) pp. 250–68. Nicholas Brealy, London.

Kets de Vries, M.F.R. (2000). *Happiness: A Guideline* (in Dutch). Nieuwerzijds, Amsterdam.

Kets de Vries, M.F.R. (2002). *The Happiness Equation*. Vermilion, London.

Kitching, R. (2005). Giving the water of life. *New Civil Engineer International*. April, 19–20.

Klein, N. (2002). *No Logo*. Picador, New York.

Knill J. (2003). Core values: the first Hans Cloos lecture. *Bulletin of Engineering Geology and the Environment*, **62**(1), 1–34.

Knights, M. (2005). *Risk Management of Tunneling Works*. Presentation to Advisory Board, 23 November. Hogeschool Zeeland, Vlissingen.

Koelewijn, A.R. (2002). The practical value of slope stability models. In *Learned and Applied Soil Mechanics out of Delft* (F.B.J Barends and P.M.P.C.Steijger, eds) pp. 107–14. A.A. Balkema, Lisse.

Koenen, I. (2004). Large project disappointments are not necessary: effective risk allocation by model contracts (in Dutch). *Cobouw*, 15 November, 1.

Korff, M. (2003). Crossing of bored tunnel with existing railway: risk for liquefaction controlled. In *Proceedings Young Geotechnical Engineers Conference*, September, Romania.

Kort, D.A. (2002). *Sheet Pile Walls in Soft Soil*. PhD Thesis, Delft University Press, Delft.

Lacasse, S. and Nadim, F. (1998). Risk and reliability in geotechnical engineering. In *Proceedings Fourth International Conference on Case Histories in Geotechnical Engineering*, St Louis, Missouri, March 9–12, pp. 1172–92. University of Missouri-Rolla, Rolla.

Laverman, W. (2005). It takes time to become a good client (in Dutch).*Building Business*, **1**, 28–31.

Leifer, R. (1999). *The Happiness Project*. Snow Lion Publications, New York.

Letza, S.R. (2000). *Syllabus Crisis Management*, NIMBAS-Bradford MBA Programme, September 2000. NIMBAS University, Utrecht.

Lowery, C.M., Beadles, N.A. and Carpenter, J.B. (2000) TQM's human resource component. *Quality Progress*, February, 55–9.

Luger, H.J., van den Hoek, E.E., van Tol, A.F. (2003). Souterrain The Hague: clogging of groundwater wells above a gel layer during construction of an underground tram station. In *Proceedings ITA World Tunnelling Congress*, April, Amsterdam, pp. 387–93. Balkema, Lisse.

Lunne, T., Roberston, P.K. and Powell, J.J.M. (1997). *Cone Penetration Testing in Geotechnical Practice*. Spon Press, London.

Macpherson, F. (2001). Risky business. *International Construction*, **40**(3), 41.

Marx Hubbard, B. (2004) *Conscious Evolution: Awakening the Power of our Social Potential*, Dutch edn. Aionion Symbolon, Amstelveen.

Matyas, R.M., Mathews, A.A., Smith, R.J. and Sperry, P.E. (1996). *Construction Dispute Review Board Manual*. McGraw-Hill, New York.

Martin, D. (1987). Dry run for Washington Metro gives NATM an American boost. *Tunnels and Tunneling*, **19**(5), 16–8.

Mining Communications (2004). *No-Dig World News: Risk Management Code of Practice*, 5 January. Mining Communications, London.

Mintzberg, H. (1998). The professional organisation. In *The Strategy Process – Revised European Edition* (H. Mintzberg, J.B. Quinn and S. Goshal, eds) pp. 681–96. Pearson Education, Harlow, Essex.

Mintzberg, H., Quinn, J.B. and Goshal, S. (1998). *The Strategy Process – Revised European Edition*. Pearson Education, Harlow, Essex.

Molendijk, W.O. and Aantjes, A.T. (2003). Risk management of earthworks using GeoQ. In *Proceedings 22nd World Road Congress*, October, Durban. PIARC, Paris.

Molendijk, W.O. and van den Berg, F.P.W. (2003). Well-founded risk management for the Betuweroute freight railway by the observational method. In *Proceedings of XIIIth European Conference on Soil Mechanics and Geotechnical Engineering*, August, Prague, pp. 741–46. CGTS, Prague.

Molendijk, W.O., Van, M.A. and Dykstra, C.J. (2003). Improved reliability of (rest) settlement predictions of embankments on soft soils. In *Proceedings 22nd World Road Congress*, October, Durban. PIARC, Paris.

Mora, S. and Keipi, K. (2005). Incorporation of engineering geological modeling in disaster risk management on development investment projects. In *Proceedings of the Symposium New Developments in Geo Information Handling for Engineering Geology*, 21 January 2005 on the occasion of the retirement of IAEG President Niek Rengers, Abstact, p. 10. ITC, Enschede.

Morgenstern, N.R. (2000a). Common ground. In *Proceedings International Conference on Geotechnical and Geological Engineering*, 19–24 November, Melbourne, Australia, Vol. 1, pp. 1–30. Technomic, Basel.

Morgenstern N.R. (2000b). *Performance in Geotechnical Practice*. The Inaugural Lumb Lecture, 10 May. Hong Kong.

Mott McDonald and Soil Mechanics (1994). *Study of the Efficiency of Site Investigation Practices*. Project report 60. Transport Research Laboratory, Crowthorne.

Muhlemann, A.P., Oakland, J.S. and Lockyer, K.G. (1992). *Production and Operations Management*, 4th edn. Pitman Publishing, London.

Mulhearn, C. and Vane, H.R. (1999). *Economics*. Macmillian Press, London.

Munfah, N., Zlatanic, S. and Baraclough, P. (2004). Managing underground construction risk in New York. In *North American Tunneling 2004* (Ozdemir, ed.) pp. 163–70. Taylor & Francis Group, London.

Naisbitt, J. (1996). From nations to networks. In *Rethinking the Future* (R. Gibson, ed.) pp. 212–27. Nicholas Brealy Publishers, London.

Naisbitt, J. (1984). *Megatrends: Ten New Directions Transforming our Lives*. Warner Books, New York.

Network for Contaminated Land in Europe (2002). *Cost-Effective Site Characterisation: Dealing with Uncertainties, Innovation, Legislation Constraints*. Conclusions of the NICOLE workshop, 18–19 April (P. Bardos, ed.). NICOLE, Pisa.

Neff, T.L. (1998). Risk management considerations for complex subsurface projects. In *Subsurface Conditions: Risk Management for Design and Construction Management Professionals* (D.J. Hatem, ed.) pp. 129–61. John Wiley & Sons, New York.

NEN 6740 (1991). *Geotechnics 1990 Basic Requirements and Loads* (in Dutch). Netherlands Standardisation Institute, Delft.

Ni, H.C. (1997). *Entering the Tao*. Shambala Publications, Boston.

NN (2005) Nun stops tunnel. *International Construction*, **44**(2), 3.

Nußbaumer, M. and Nübel, K. (2005). Portfolio based approach to project risk management. In *Proceedings 16th International Conference on Soil Mechanics and Geotechnical Engineering*, 12–16 September, Osaka, Japan, pp. 2833–36. Millpress, Rotterdam.

Oakland, J.S. (1993). *Total Quality Management*, 2nd revised edn. Butterworth-Heinemann, Oxford.

O'Connell, S. (1997). *Mindreading*. William Heinemann, London.

Ortoli, S. and Pharabod, J.P. (1986). *The Soluble Fish and Other Paradoxes: The Debate about Quantum Theory* (Dutch edn). Van Gennip, Amsterdam.

Paine, L., Deshpandé, R., Margolis, J.D. and Bettcher, K.E. (2005). Up to code: Does your company's conduct meet world-class standards? *Harvard Business Review*, December, 122–33.

Peck, R.B. (1969). Advantages and limitations of the observational method in applied soil mechanics. *Géotechnique*, **19**, 171–87.

Pereboom, D.P., Tiggelman, L. and van Staveren, M.Th. (2005). Geotechnical risk management in The Netherlands. In *Proceedings 16th International Conference on Soil Mechanics and Geotechnical Engineering*, 12–16 September, Osaka, Japan, pp. 2841–4. Millpress, Rotterdam.

Perry, P.E. (1981). Evaluation of savings for underground construction. *Underground Space*, **6**, 29–42.

Peters, T.J. and Waterman, R.H. (1982). *In Search of Excellence*. Harper & Row Publishers, New York.

Piercy, N. (1997). *Market-Led Strategic Change*, 2nd edn. Butterworth-Heinemann, Oxford.

Portugal, J.C., Portugal A., and Santo, A. (2005). Excavation induced damage. In *Proceedings 16th International Conference on Soil Mechanics and Geotechnical Engineering*, 12–16 September, Osaka, Japan, pp. 1543–6. Millpress, Rotterdam.

Poulos, S.J. (1998). Role of the professional consultant in the evaluation of differing site condition claims. In *Subsurface Conditions: Risk Management for Design and Construction Management Profesionals* (D.J. Hatem, ed.) pp. 211–22. John Wiley & Sons, New York.

Press, F. and Siever, R. (1982). *Earth*, 3rd edn. W.H. Freeman and Company, San Francisco.

Priemus, H. (2005) Murphy's law – design and construct at the high speed railway link (in Dutch). *Building Business*, February, 42–9.

Process and System Innovation in Building and Construction (2004). *Inventory of International Reforms in Building and Construction*. Report PSIB017_S_04_2341, June. PSIBouw, Gouda.

Quint, M. (2005). Talking point: so what exactly is 'contaminated land' then? *Ground Engineering*, November, 7.

Reason, J. (1990). *Human Error*. Cambridge University Press, Cambridge.

Rigby, P. (1999). Identifying and managing ground risks. *Tunnels and Tunneling International*, October, 39–43.

Rohrmann, B. (1998). The risk notion: epistomological and empirical consideration. In *Integrated Risk Assessment* (M.G. Stewart and R.E. Melchers, eds) pp. 39–46. Balkema, Rotterdam.

Roobeek, A., Mandersloot, E. and du Marchie Servaas, C. (1998). Organizational transformation through strategic dialogue: The Schiphol case. *Concepts and Transformation: International Journal of Action Research and Organizational Renewal*, **3**, 53–77.

Sassen, S. (2004). *The New Intermediate Economy: Between Uncertainty and the Client*. Third Ernst Heijmans Lecture, 18 November. www.ernstheijmanslezing.nl, Utrecht.

Savadis, S.A. and Rackwitz, F. (2004). Geotechnics and environment: A consideration by planning and construction of the transport infrastructure in the centre of Berlin. In *Proceedings: Fifth International Conference on Case Histories in Geotechnical Engineering*, 13–17 April, New York (S. Prakash, ed.). University of Missouri-Rolla, Rolla.

SBR (2000). *Risk Management is Profit Management: A Guideline for Small and Medium Enterprises* (in Dutch). Stichting Bouw Research, Rotterdam.

Schaberg, J. (2005). Infrastructure and financial returns (in Dutch). *NRC*, 21 July, 9.

Schein, E.H. (1984). Coming to a new awareness of organisational culture. *Sloan Management Review*, Winter, 3–16.

Schon, D. (1983). *The Reflective Practioner*. Basic Books, New York.

Seligman, M.E.P. (2002). *Authentic Happiness: Using the New Positive Psychology to Realize your Potential for Lasting Fulfillment*. Free Press, London.

Senge, P.M. (1990). *The Fifth Discipline: The Art and Practice of the Learning Organisation*. Doubleday, New York.

Shrivastava, P., Mitroff, I.I., Miller, D. and Miglani, A. (1988). Understanding industrial crises. *Journal of Management Studies*, **25**, 283–303.

Skipp B.O. (ed.) (1993). *Risk and Reliability in Ground Engineering*. Thomas Telford, London.

Sleight, C. (ed.) (2005a). Construction growth expected to increase. *International Construction*, January-February, 12–4.

Sleight C. (ed.) (2005b). IC comment. *International Construction*, July-August, 3.

Slovic, P., Fischoff, B. and Lichtenstein S. (1982). Why study risk perception? *Risk Analysis*, **2**, 83–93.

Smallman, C. (2000). *Crisis and Risk Management*. Lecture Presentation, NIMBAS-Bradford MBA Programme, September 2000. NIMBAS University, Utrecht.

Smallman, C. (1999). The Risk Factor. *Financial Times*, 30 December, 42–5.

Smallman, C. (1998). *Risk Perception: State of the Art*. University of Bradford Management Centre Working Paper No. 9820. University of Bradford Management Centre, Bradford.

Smallman, C. (1996). Challenging the orthodoxy in risk management. *Safety Science*, **22**(1–3), 245–62.

Smith, N. (2003). *Appraisal, Risk and Uncertainty*. Thomas Telford, London.

Smith, N. (ed.) (1998). *Managing Risk in Construction Projects*. Blackwell Science, Oxford.

Smith, R. (1996). Allocation of risk: the case for manageability. *The International Construction Law Review*, **4**, 549–69.

Smith, R.J. (1999). Identifying and allocating risk. In *Construction Law Handbook* (R.F. Cushman, ed.) pp. 459–97. Aspen Publising, New York.

Soudain, M. (2000). Talking about a revolution. *Ground Engineering*, September, 12–3.

Sperry, P.E. (1981). Evaluation of savings for underground construction. *Underground Space*, **6**, 29–42.

Tani, K. (2001). *Allocation of Risk Associated with Ground Conditions in Privatised Iinfrastructure Contracts*. MSc Thesis. Department of Construction Management & Engineering, University of Reading, Reading.

Thompson, J.L. (1997). *Strategic Management: Awareness and Change*, 3rd edn. International Thomson Business Press, London.

Thompson, P. and Perry, J. (1992). *Engineering Construction Risks*. Thomas Telford, London.

Toft, B. (1993). Behavioural aspects of risk management. In *Proceedings Annual Conference of Association of Risk Managers in Industry and Commerce*, 1–4 April, University of Warwick, Warwick.

Toft, B. (1996) Limits to the mathematical models of disasters. In *Accident and Design: Contemporary Debates in Risk Management* (C. Hood and D. Jones, eds.) pp. 99–110. UCL Press, London.

Transparency International (2005). *Global Corruption Report 2005*. T.I. Pluto Press, London.

Trout, J. and Rivkin, S. (2000). *Differentiate or Die: Survival in our Era of Killer Competition*. John Wiley & Sons, New York.

Turner, C. (1995). *The Eureka Principle*. Element Books, Shaftesbury.

Turner, D.F. and Turner, A. (1999). *Building Contract Claims and Disputes*, 2nd edn. Longman, London.

Uhlfelder, H.F. (2000). It's all about improving performance. *Quality Progress*, February, 47–52.

US Army Corps Of Engineers (1989). *Partnering – Alternative Dispute Resolutions Series Pamphlet*. No. #4/TWR Pamphlet-91ADR-P-4. COE, Washington DC.

US Subcommittee on Geotechnical Site Investigations (1984). *Geotechnical Site Investigations for Underground Projects – Volume 1: Overview of Practice and Legal Issues, Evaluation of Cases, Conclusions and Recommendations*. US National Academic Press, Washington DC.

van Meurs, G.A.M., de Cleen, M.P.T.M. Taat, J. and Schurink, E. (2001). Flexible emission control: a process-like approach. In *Proceedings Seminar on Analysis, Methodology of Treatment and Remediation of Contaminated Soils and Groundwater*, 13–15 March. United Nations, Economic Commission for Europe, Paris.

van Meurs, G.A.M., van Ree, C.C.D.F., van de Velde, J.L. and Oosterom, W. (2001). Flexible and dynamic site investigation: faster, better and cheaper. In *Field Screening Europe 2001* (W. Breh et al., eds.) pp. 113–7. Kluwer Academic Publisers, Deventer.

van Oirschot, R. (2003). *Future Management: The Paradox of Future Control* (in Dutch). Business Contact, Amsterdam.

van Staveren, M.Th. (2005). Willingness to see, be and change. In *Change Management: Current Visions on Management* (in Dutch), pp. 215–32. SDU, The Hague.

van Staveren, M.Th. (2004). Soil investigations and contractual risk alllocation. In *Syllabus of the International Site Investigation Course*. GeoDelft Academy, Delft.

van Staveren, M.Th. (2001a). The relation between monitoring, risk allocation and geotechnical baseline reports. In *Syllabus International Course on Geotechnical Instrumentation for Field Measurements*. GeoDelft, Delft.

van Staveren, M.Th. (2001b). *Divestment or Leadership?* MBA Thesis of NIMBAS-Bradford MBA Programme. NIMBAS University, Utrecht.

van Staveren, M.Th. and Bolijn, J. (2003). Early attention to geotechnical risk provides interesting options (in Dutch). *Geotechniek*, **7**(2), 12–7.

van Staveren, M.Th. and van Deen, J.K. (1998). The need for cone penetration test accuracy classes. In *Proceedings Eighth International Congress International Association for Engineering Geology and the Environment*, 21–25 September, Vancouver, Canada (D. Moore and O. Hungr, eds) pp. 259–65. Balkema, Rotterdam.

van Staveren, M.Th. and Knoeff J.G. (2004). The Geotechnical Baseline Report as risk allocation tool. In *Proceedings EurEnGeo 2004*, May, Liege, pp. 777–85. Springer Verlag, Berlin.

van Staveren, M.Th. and Litjens, P.P.T. (2001). Risk allocation by the Geotechnical Baseline Report (in Dutch). *Land en Water*, **41**(1/2), 46–7.

van Staveren, M.Th. and Peters T.J.M. (2004). Matching monitoring, risk allocation and Geotechnical Baseline Reports. In *Proceedings EurEnGeo 2004*, May, Liege, pp. 786–91, Springer Verlag, Berlin.

van Staveren, M.Th. and van Seters A. (2004). Smart site investigations save money! In *Proceedings EurEnGeo 2004*, May, Liege, pp. 792–800. Springer Verlag, Berlin.

Vattimo, G. (1992). *The Transparent Society*. Polity Press, Cambridge.

Versluis, J. (1995). Better innovations by risk diagnosing (in Dutch). *De Ingenieur*, **12**, 33–5.

Versteegen, J. and van Staveren, M.Th. (2004). Missed opportunities by neglecting uncertainty (in Dutch). *Cobouw*, 17 December, 3.

Viehöfer, Th.C. (2002). *Tools and Methods for Geotechnical Risk Analysis as Part of Risk Management within the RISMAN Rramework* (in Dutch). Report 510160.0001. GeoDelft, Delft.

Waring, A.E. (1996). *Safety Management Systems*. Chapman & Hall/ITBP, London.

Waring A.E. and Glendon, A.I. (1998), *Managing Risk*. International Thomson Business Press, London.

Watzlawick, P., Beavin, J.H. and Jackson, D.D. (1967). *Pragmatics of Human Communications*. W.W. Norton & Company, New York.

Weatherhead, M., Owen, K. and Hall, C. (2005). *Integrating Value and Risk in Construction*. CIRIA, London.

Wentink, J.J. (2001). *Building E-visualisations* (in Dutch). Systeem Drukkers, Delft.

Wijffels, H. (2004). Foreword. In *Conscious Evolution: Awakening the Power of our Social Potential*, Dutch edn. (B. Marx Hubbard, ed.) pp. 12–4. Aionion Symbolon, Amstelveen.

Wildavsky, A. and Dake, K. (1990). Theories of risk perception: who fears what and why? *Daedalus*, **199**(4), 41–60.

Wildman, W.R. (2004). Risk allocation in tunnel construction contracts. In *North American Tunneling* (Ozdemir, ed.) pp. 171–6. Taylor & Francis Group, London.

Witten. D. and Tulku Rinpoche, A. (1998). Enlightened management. Random House, London.

Wuite, M. (2005). Alliance contract deserves follow-up (in Dutch). *Cobouw*, 26 October, 7.

Yin, J.H. and Graham, J. (1999). Elastic viscoplastic modelling of time-dependent stress-strain behaviour of soils. *Canadian Geotechnical Journal*, **36**, 736–45.

Zohar, D. and Marshall, I. (2004). *Spiritual Capital*. Bloomsbury, London.

Index